CARBON DIOXIDE THROUGH THE AGES

HAN DOLMAN

CARBON DIOXIDE THROUGH THE AGES

From wild spirit to climate culprit

OXFORD
UNIVERSITY PRESS

Great Clarendon Street, Oxford, OX2 6DP,
United Kingdom

Oxford University Press is a department of the University of Oxford.
It furthers the University's objective of excellence in research, scholarship,
and education by publishing worldwide. Oxford is a registered trade mark of
Oxford University Press in the UK and in certain other countries

Published in the United States of America by Oxford University Press
198 Madison Avenue, New York, NY 10016, United States of America

British Library Cataloguing in Publication Data
Data available

Library of Congress Control Number: 2022949984

ISBN 978–0–19–886941–2

DOI: 10.1093/oso/9780198869412.001.0001

Printed and bound by
CPI Group (UK) Ltd, Croydon, CR0 4YY

PREFACE

In May 2022 the global average concentration of CO_2 in the atmosphere hit 420 ppm, 50% more than the pre-industrial value of 280 ppm. The level of carbon dioxide had not been so high on our planet for 4 million years. In early June 2022 I spent a day at the plenary meeting of the Subsidiary Body for Scientific and Technological Advice of the UN Convention on Climate Change in Bonn, Germany where almost everything related to the relentless rise in CO_2 came across in the meeting: responsibility for historical emissions, adaptation to climate change, equity, social welfare, and regulations to stop the emissions. These two events, with hindsight, together provide almost a complete summary of the rationale for writing this book.

In 2015 the Paris Agreement had been put forward as a successor to the ailing Kyoto Protocol. It aimed to set limits of climate change to 2 °C or preferably 1.5 °C. This was widely considered a landmark agreement. A few years later I realized that the perception of CO_2 in society had changed. It was no longer a largely unknown, harmless substance, instead it had become the climate culprit *par excellence*. I began to wonder why and how this had changed (and why not earlier) and came to realize that an investigation into the history of CO_2 might help in providing answers. This book is the result of that investigation. It inevitably contains biases. I am a natural (earth) scientist with an interest in the history and sociology of science, but not an expert in those latter fields. Nevertheless,

I hope that the book achieves its aims to trace CO_2 through the ages and geological time, explaining its role in the atmosphere, ocean, and land and finally the impact of human burning of fossil fuel and deforestation. This story would not be complete if I did not include the political process that led to the Paris Agreement, through increased scientific awareness first, and a series of large international conferences and meetings that put climate change on the political agenda. It also deals with the lack of urgency of governments, that have for too long blissfully ignored the scientific evidence that was provided to them by organizations like the Intergovernmental Panel on Climate Change.

The whole book has been carefully read by Ingeborg Levin and Gerald Ganssen who have, as always, identified sloppy writing and provided new perspectives on the matter. Parts of the book have been read by Guido van der Werf, Joost Tielbeke, Rob Buiter, Henk Brinkhuis, Matthew Humphreys, Anja Spang, Kim Sauter, and Pierre Offre. The first versions have been edited by John Gash, who not only gave valuable advice on the writing but often also on its content and provided a continuous stimulus to go on. Zicarlo van Aalderen provided invaluable help in ordering the notes and with establishing the copyright of some of the figures. Quite a few of those were (re)drawn by Barbara Pillip. I am greatly indebted to all these people, although any remaining errors remain of course my own responsibility. I am grateful to my editor at Oxford University Press, Adlung Sonke, who quite quickly took to the idea of this book, and made it possible.

Writing this book was interrupted by a pandemic and a change of jobs, from a university professor in the department of earth sciences at the Vrije Universiteit of Amsterdam to the Royal Netherlands Institute for Sea Research. This caused some delay in the writing of the final chapters, but also widened my perspective. The publishing of three IPCC reports and the increasing frustration among my younger scientist colleagues about the lack of urgency among policymakers provided the final push for me to finish the book.

The support of my partner Agnes and my sons Jim and Wouter during the writing of the book was invaluable, as it is for everything else in my life.

CONTENTS

CHAPTER I

CARBON DIOXIDE, FROM A WILD SPIRIT TO CLIMATE CULPRIT

*T*his book traces the history of carbon dioxide through the ages. Carbon dioxide made a renewed star appearance on the world's stage in 2015 when the governments of the world agreed in Paris to limit global warming to a maximum of 2 °C—but preferably 1.5 °C compared to pre-industrial time (around 1850). This difference might seem small, but the higher the temperature, the more the planet will suffer from extreme events: droughts, extreme flooding, rise in sea level, heatwaves, and loss of biodiversity. So, something needs to be done, we need to reduce our emissions of carbon dioxide fast, incredibly fast.

The book traces the development of the perception of CO_2 through the ages. We discuss its role in climate through the absorption of infrared radiation. It deals with its discovery by the Flemish doctor van Helmont, to its final chemical understanding, its role in photosynthesis and respiration, the variability through geological time, and closer in time to us, variability during the ice ages. We then enter the period when carbon dioxide becomes a human phenomenon though fossil fuel burning and deforestation. It describes the world's efforts to control the rise in CO_2 through now more than 60 years, since it became obvious that humans were changing the composition of the atmosphere and climate. We then discuss what needs to be done to stop *climate from warming beyond the limits set in the Paris Agreement. It can be done, but it is tough…*

The world watched in anxiety in December 2015 when representatives of the 194 countries of the United Nations convened in Paris to negotiate a new climate agreement. The previous conference in Copenhagen in 2009 had been a thorough disappointment to the delegates of most countries. And, to the multitude of stakeholders that are involved in such meetings, such as representatives of non-governmental organizations, fossil fuel and nuclear fuel lobbyists and organizations claiming to have the ultimate solutions to the climate crisis, and not to mention the general public. There was a need to reach an agreement this time. Not only because the Kyoto protocol, that was negotiated in 1997 with help of the later Nobel prize winner Al Gore, was virtually dead and needed an update but also because the world had seen the number of extreme weather events increasing that could be directly linked to climate change. There was a sense of urgency among the delegates, shared by the local French organizers and, importantly, the world.

In the final hours of the meeting on Saturday 10 December, the final text was put forward for approval to the delegates and approved almost unanimously. The organizers of the conference, now including for the final meeting the UN Secretary General Ban Ki Moon, the French President François Hollande, and the head of the UN Climate Change Secretariat, Christiana Figueres, stood up behind the table and together raised their arms in victory to a standing ovation from the delegates that lasted several minutes. The millions of people all over the world watching the live broadcast issued a sigh of relief. They were aware of the momentous importance of the agreement. This was the first time that the UN countries really had agreed that the climate crisis needed to be halted, and in such a way that global temperature should not rise above 2 °C, or preferable above 1.5 °C.

While the difference between 1.5 °C and 2 °C may sound small, the impact on ecosystems and humans is not. In fact, with every tenth of a degree of additional warming the world is experiencing more droughts, more flooding, and more extreme weather. Take our coral reefs for instance, where a 0.5 °C difference implies[1] still a 23% occurrence at 1.5 °C warming versus an almost complete absence of corals with only 1% left, in the case of a 2 °C warming. Half a degree extra warming would imply a 2.5 times increase in extreme heat, a 10-fold increase into the occurrence of ice-free summers in the Arctic seas, a doubling of species loss, both for plants and for vertebrates, 40% more of our permafrost disappearing, and two times as big a decline in marine fishery potential. This is indeed a chilling list[2] of impacts of higher temperatures, although 'chilling' might not be the right word here. With every additional 0.5 °C of warming we will experience further increases in the intensity and frequency of hot extremes, including heatwaves, and heavy precipitation, as well as agricultural and ecological droughts in some regions. Changes in intensity and frequency of meteorological droughts, with more regions showing increases than decreases, are also predicted to occur. Increases in the frequency and intensity of hydrological droughts become larger with every tenth of a degree of global warming. There will be an increasing occurrence of some extreme

[1] IPCC, 2018
[2] IPCC, 2021–1

events unprecedented in the observational record with additional global warming, even at 1.5 °C of global warming.

Going to 3 °C or even 4 °C implies the danger of crossing thresholds or tipping points which make it difficult to go back to the original state of our environment and could lock the Earth system in a different mode for tens of thousands of years. Obvious examples here are the melting of the big ice sheets both in Greenland and in the Antarctic, of which the latter might have far reaching implications for global sea level rise. But also, biomes such as the Amazon Forest could no longer be sustained and large parts would likely switch to dry savannahs.

While this litany of potential disasters is sometimes criticized as 'alarmist' it is what the science tells us may happen once we cross these temperature thresholds. It tells us that we need to do everything possible to stop further warming. If that means we can stop at 1.5 °C that is good, although lower would be better; 2 °C is worse as most people would rather not live or survive in a two-degree warmer world with increased frequencies of drought in some areas and flooding in others, let alone increases in heatwaves. And the Dutch and other people on small island states and low-lying areas in the world have an in-built fear against rising sea levels.

To remain below 1.5 °C, the emission of gases that cause the planet to warm up need to be reduced to ultimately zero (and in fact below zero, see Chapter 12). This gargantuan task lies ahead of us. But what are these gases that cause climate change? Among the three most important ones, carbon dioxide stands out, the other two being methane and nitrous oxide. Carbon dioxide is essentially a waste product of the burning of fossil fuel such as oil and gas. It is, however, also the main resource for plants that convert it back into oxygen under the influence of light in a process called photosynthesis, producing sugars for growth. It was discovered in the late seventeenth century by a Belgian scientist, van Helmont, by burning charcoal and who called it a wild spirit, '*spiritus sylvester*'. Joseph Priestley later expounded on this discovery and went on to further describe its qualities in the eighteenth century. He also noted its colourless properties and established that it was produced by the combustion of materials containing carbon. How this colourless gas became so important that a new UN secretariat was formed in 1992 that led to the 2015 Paris Climate agreement is the main theme of this book. To trace that story, we need to dive into the early history of its discovery when neither the qualities nor role of carbon dioxide and oxygen were known. We enter the period of science practice at the early enlightenment. From there we move to the nineteenth century where the role of atmospheric absorption in the Earth's energy budget becomes more evident, to the important realization in the early twentieth century, that indeed carbon dioxide and climate change are intimately linked.

To understand the role of carbon dioxide on Earth we need to understand the role of the geology and ocean as the great regulator of past and present atmospheric concentrations and identify the variability in the geological past. But we also need to see how it can be measured. The efforts to monitor carbon dioxide in the air with new instrumentation, starting in 1956 by Charles Keeling, provide another milestone in our quest to understand the role of the gas in the Earth's climate. This new instrumental capability would also help with providing a record of carbon dioxide variation to almost a million years back when, for instance, tiny capsules of air that get locked up into the snow and ice of the big ice sheets of Greenland

and Antarctica are retrieved by coring deep into the ice, and are analysed for their concentration of various gases.

From there on the story becomes part science, part political. Science, to further elucidate the precise role of carbon dioxide in the atmosphere and to appreciate the role of humans, land, and ocean in controlling atmospheric concentrations. Politics, to the set-up of the Intergovernmental Panel on Climate Change and the United Nations Framework Convention on Climate Change. To the long and tedious process of making the world agree that burning of fossil fuels was indeed the main culprit of climate change and that, if we continue along that path, climate would change even further. And ultimately to the various negotiations that led to the Paris agreement.

What happened in that political process in the lead up to the Paris agreement is important, as are the numerous conferences that followed and in which the world seemed and still seems to lack the sense of urgency to implement the agreement. This has led to an almost complete disconnect between the science, individual scientists and activists that do realize the urgency of the climate crisis, and, in contrast, the global policy makers that appear to act too slowly.

This then is the story of a gas discovered in the seventeenth century that rose to political prominence in the early twenty-first century and on which the future of the next generations of humankind depends.

CHAPTER 2

WHAT DOES CARBON DIOXIDE DO IN THE ATMOSPHERE?

W e tell the story of how a tiny molecule manages to absorb radiation and thereby affects the temperature of the whole planet. Radiation comes in two sorts, that part of the spectrum (like a rainbow) that we can see, and that part of the spectrum that we cannot see. The astronomer Herschel around 1800 made several important experiments: these revealed that radiation beyond the visible part of the spectrum contained heat. When he used a special glass to spread the rays over a table set up with thermometers, he noted that they still got warm in some parts, even though he could not see that 'invisible light'. This was later called infrared radiation. His successors Maxwell and Planck identified radiation as being electromagnetic (including our well-known radio and microwaves) and further quantified the important relation between wavelength and energy. Thanks to them we now know that infrared radiation coming from the Earth has a longer wavelength and carries less energy than visible radiation coming from the Sun.

Joseph Fourier realized that the Earth's temperature resulted from a balance between different sources of radiation, that coming from the Sun and that going out from the Earth. While he made some mistakes in his 1824 calculations, his idea was the starting point for our present climate models.

In 1856 the pioneering American female scientist Eunice Foote discovered that different gases, such as CO_2, have an impact on the temperature of the air containing the gases when put under different forms of radiation. This is the birth of the greenhouse effect, usually attributed to John Tyndall. To do credit to both, John Tyndall's apparatus was a bit more sophisticated so that he could clearly show that the increase in temperature was related to Herschel's infrared radiation. The greenhouse effect was experimentally determined, at least in the laboratory.

On stage then comes the Swede Svante Arrhenius, in 1896. He was at first interested in trying to explain the differences in temperature between an ice age (a period where large parts of the Northern Hemisphere were covered by ice sheets and glaciers) and the periods in between with less ice. Geologists had started to realize that these occurred at regular intervals, but what caused these alterations between warm and cold periods was not known. To solve the conundrum, Arrhenius set out to do hundreds of thousands of calculations that purported to show how CO_2 might affect the absorption of radiation in the atmosphere. He concluded that small changes in the amount of CO_2 in the atmosphere could indeed have caused the changing temperatures during the ice ages. The greenhouse effect was determined—now not only in the laboratory but for the whole planet. He then went on to show that burning fossil fuels like coal and oil produced a steady increase in atmospheric CO_2, which would also increase the Earth's temperature. The enhanced greenhouse effect, or human-induced climate change, had appeared on the agenda. Guy Callendar, in 1938, improved the Arrhenius calculations and made new predictions of how much the temperature would increase if humans continued to burn fossil fuel. He was close, but not yet quite there; however, Climate Science was born.

What exactly does carbon dioxide do in the atmosphere, and why is it so important given its marginal contribution as a trace gas, just a tiny 0.04%,[1] to the composition of the atmosphere? Why does such a small amount of carbon dioxide generate such a large effect on the heat balance of the planet? And why does an increase from 0.03% to 0.04% (300 to 400 ppm) over the last two and a half centuries generate so much concern? For understanding this greenhouse effect, we must go back in time to the work of the astronomer Sir Frederick William Herschel. Although mostly famous for his discovery of the planet Uranus in 1781, he also discovered what we now call infrared radiation. The opening line of his 1800 paper is of such beauty that it is worth quoting: *'It is sometimes of great use in natural philosophy, to doubt of things that are commonly taken for granted; especially as the means of resolving any doubt, when once it is entertained, are often within our reach.'*[2] Herschel was investigating the refraction of sunlight through a prism and was interested in the question of how much heat was filtered through. He realized that the filters of different colours seemed to pass different amounts of heat. He also noted that the *'heating power of the prismatic colours, is very far from being equally divided, and that the red rays are chiefly eminent in that respect'*.[3] He also noticed that the peak of heat transfer did not coincide with the peak in radiation intensity, but rather that the measurements showed a trend, with the important possibility that this might continue beyond the red end of the visible spectrum. This radiation he called 'invisible

---

[1] The amount of CO_2 in the atmosphere is so little that it is normally expressed as parts per million (ppm), which is the number of CO_2 molecules relative to 1 million other molecules of air.

[2] Herschel, W., 1800–1

[3] See Herschel, W., 1800–1. For a good analysis of this discovery in the context of modern understanding see White, J., 2012.

light', radiation outside the visible spectrum, but still containing energy.

He went on to execute a new set of experiments to further investigate this phenomenon, which he reported in a second paper.[4] He let sunlight, after passing through a prism, fall on a table where he had marked a further five lines beyond the red part of the spectrum. He measured the temperature in these areas, using blackened thermometers, so that they would better absorb the radiation (Figure 2.1). In this second paper he was close to suggesting that infrared radiation and visible radiation were similar things, differing only in wavelength (a concept he was not aware of)

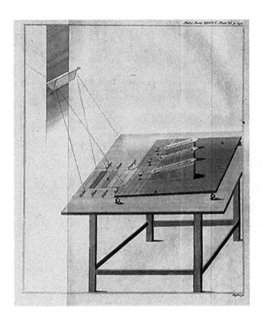

Figure 2.1 The table on which Herschel discovered infrared radiation ('heat-making rays'). Radiation falls through the prism and on the table are marked the colours, and the additional five areas after the red colour which he identified carried heat. Note also, the thermometers in this area.
Herschel, W., 1800–2

[4] Herschel, W., 1800–2

and energy. He was close enough: *To conclude, if we call light, those rays which illuminate objects, and radiant heat, those which heat bodies, it may be inquired, whether light be essentially different from radiant heat? In answer to which I would suggest, that we are not allowed, by the rules of philosophizing, to admit two different causes to explain certain effects, if they may be accounted for by one'.* If only he had stopped there…

But he went on investigating radiation, published a third and fourth paper and finally concluded that they were not of similar nature, that the heat-containing rays were different from the ones carrying light (i.e. the visible rays). After several hundred (!) new experiments he produced a graph in his final paper of 6 November 1800 that contained the overlapping spectral distributions of visible light and infrared radiation. He called these curves the 'spectrum of illumination' and the 'spectrum of heat' and unfortunately drew the wrong conclusion: *'A mere inspection of the two figures … will enable us now to see how very differently the disperses the heat-making rays, and those which occasion illumination'.* What exactly restrained him from arriving at the correct conclusion remains somewhat unclear—it could be that the way he arrived at his last graph was problematical and wrong as some authors suggest[5]. He certainly lacked the conceptual understanding of radiation that would later help Maxwell unify the spectrum. He is remembered rightly though, for being the first to have discovered rays carrying heat, or radiation outside visible light and having noted that they behaved in a somewhat similar way to visible light. A year later the German scientist Ritter would discover the 'invisible' rays existing beyond the other, blue end of the spectrum, ultraviolet radiation.

[5] White, 2012.

The Scot James Clerk Maxwell took this work an important step further. His greatest contribution was published in 1865 but had been in the making for much longer. Maxwell had the ability to return to his previous work and further develop the concepts in an uncanny way.[6] His 1865 paper 'A *dynamical theory of the electromagnetic field*'[7] is one of the classics of scientific literature. In more than 50 pages and a multitude of equations, he showed how the forces of electricity and magnetism could be put into one set of equations, known as the Maxwell equations. Every problem related to magnetism and electricity could in principle be solved by these equations as he carefully showed in his paper. In section VI, called the electromagnetic theory of light, he sets out to '*investigate whether these properties of that which constitutes the electromagnetic field, deduced from electromagnetic phenomena alone, are sufficient to explain the propagation of light through the same substance*'. He develops the equations and arrives at the conclusion that these waves travel at the speed of light: '*the agreement of the results seems to show that light and magnetism are affectations of the same substance, and that light is an electromagnetic disturbance propagated through the field according to electromagnetic laws*'. He compared the results of his equations, which suggest a value for the speed of light very close to those observed at the time. He had shown that visible radiation was an electromagnetic wave, travelling at the speed of light. He also suggested that other, non-visible radiation and radiant heat (remember Herschel) were in fact electromagnetic waves. It would take the experimental work of the German physicist Heinrich Hertz to confirm that Maxwell's theory of electromagnetic waves was indeed correct. He showed that waves at wavelengths

---

[6] Mahon, B., 2003
[7] Maxwell, J., 1865

now known as radio waves behaved in a similar way to light and could be reflected and refracted just as visible light. Who deserves better the title of discoverer of electromagnetic waves: the theorist James Clerk Maxwell or the experimenter Heinrich Hertz? It is an open question and probably better left to historians of science.

A few years before Maxwell developed his equations, the Irishman John Tyndall performed a set of experiments that investigated the effect of several gases on the transmission of radiation. These form the basis of our current understanding of the greenhouse effect. Coming from investigating glaciers he was drawn to the question of how solar and terrestrial heat were transmitted through the Earth's atmosphere, stimulated by the work of de Saussure and Fourier. De Saussure had developed a so-called heliothermometer, which consisted of a wooden box, insulated with cork and wool, but having a glass lid. The inside of the box was painted black so that it would absorb all radiation. A thermometer inside the box revealed how warming was related to the amount of incoming radiation. De Saussure developed the box to test the hypothesis that it was cooler at the top of a mountain because the radiation was less intense. He was unable to prove this: in fact the decline in temperature with height is due to the fact that air expands and cools when it is lifted (the so-called dry adiabatic lapse rate predicts a rate of change of temperature with height of about 10°C per kilometre).

Although Joseph Fourier is sometimes referred to as the discoverer of the greenhouse effect,[8] this is incorrect: he never once mentions the greenhouse effect in his original paper entitled 'Remarques general sur les températures du globe terrestre des espaces

---

[8] Fleming, J., 1998

planétaires' that is mostly cited in this context. This paper was published in 1824, but a similar paper that was published in 1827 is more widely read and referred to. Its title translates as, '*On the temperature of the terrestrial sphere and interplanetary space*'.[9] Those papers also do not mention the greenhouse effect. Why Fourier is mentioned as one of the first people studying the greenhouse effect is that he arguably was one of the first to study the problem of the temperature of the Earth in the context of solar radiation. His key contribution was that he set out to investigate the balance between three sources of heat. He states that '*i) The Earth is heated by solar radiation, the unequal distribution of which produces the diversity of climates; ii) it participates in the common temperature of interplanetary space, being exposed to the radiation by countless stars which surround all parts of the solar system and iii) the Earth has conserved in the interior of its mass, a part of the primordial heat which it had when the planets were originally formed.*' By posing the problem of the Earth's temperature as the result of the incoming and outgoing heat, he basically set the problem in a way that still forms the basis of our current climate models. He was, however, somewhat wrong with his second assumption in which he stated that the Earth receives radiation in the form of heat from the interplanetary or interstellar space. The interstellar space is now known to be close to the absolute zero of temperature (2.7 K or −270 °C; absolute zero is defined as 0K or −273.15 °C) and hence does not contribute much heat to the Earth.

---

[9] See Fleming, J., 1998 for an extensive discussion of the publication and citation history of this paper. 'On the temperature of the terrestrial sphere and interplanetary space' is the translation of 'Remarques general sur les températures du globe terrestre des espaces planétaires'. Fourier, J.-B., 1824. Translated by Pierrehumbert, R., 2004. Available in Archer, D. & Pierrehumbert, R., 2011, a collection of key papers that should be on the shelves of every climate scientist.

What he was not wrong about was the difference in absorption capacity of visible light and dark heat, as he called the invisible rays carrying heat detected by Herschel (or as we would say, infrared radiation). When he describes de Saussure's box, and uses it as an analogy for the heating of the Earth's atmosphere, he states that *'the heat emanated by the Sun has properties different from those of dark heat'* and continues further down: *'because the heat has less trouble penetrating the air when it is in the form of light, than it has when exiting back through the air after it has been converted to dark heat'*. The final quote is as close to describing the greenhouse effect as Fourier gets, without mentioning it explicitly.

Back to Tyndall. At this time, Tyndall was a towering figure in science in Britain and played a leading role in the Royal Society. When he published his 1861 Bakerian lecture, his second, he mentions no less than eight memberships of scientific societies abroad. By the end of his career, he would count 35. Trained as a surveyor and cartographer at the British Ordnance Survey, he obtained his PhD in Marburg, Germany and was elected Fellow of the Royal Society in 1852. In 1853 he was granted a professorship at the Royal Institution,[10] where in 1867 after Faraday's death he became his successor as the director. When he started his work on the absorption of trace gases, he was convinced that chemical structure also mattered—though he could obviously not have been aware of the detailed structure of the molecules and atoms he investigated. In his paper he phrases it as such: *'But in the cases above recorded the molecules are perfectly free, and we fix upon them individually the effects which the experiments exhibit. Thus the mind's eye is directed more firmly than ever on those distinctive physical qualities whereby*

[10] Jackson, R., 2018.

a ray of heat is stopped by one molecule and unimpeded by another.[11] Tyndall was interested in the interaction of radiation and molecular structure, and most likely saw himself as less of a climate scientist than history has made him to be.

The apparatus Tyndall built to investigate the effect of gases on the absorption of infrared rays, was both clever and unique. It comprised a large tube which was filled with gas (Figure 2.2). The real innovation of this differential spectrometer was that it did not try to measure the absolute impact of gas on the transmission of thermal radiation, but that it used a differential measurement—this greatly increased the sensitivity. A thermopile (a series of thermocouples that are wired together producing a voltage dependent on the difference between the temperatures of the hot and cold junctions) then measured the difference between the radiation coming from the cube of boiling water only, and that coming from the cube of boiling water at the other end after transmission through the gas. This innovation allowed Tyndall to reach much higher accuracy in measurements of the effect on radiation at

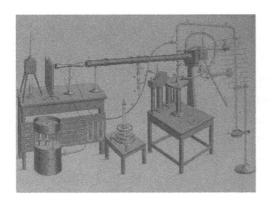

Figure 2.2 Tyndall's experimental set-up to investigate the effect of thermal radiation coming from cubes c on the thermopiles located at p.
Tyndall, J., 1861

---

[11] Tyndall, J., 1861. See Jackson, R., 2018.

much lower concentrations of the gases. Without this experimental design he would have been unable to detect the effects of small concentrations or density on the transmission.

But Tyndall was not the first! The honour of discovering that atmospheric gases can absorb radiation and thereby cause an increase in temperature goes to the American scientist and women's rights campaigner Eunice Foote.[12] Here again, we see scientific developments taking place independently, almost at the same time, in different parts of the world. However, Eunice Foote's contribution somehow got lost in history, and was only recently rediscovered. Her paper, *'Circumstances affecting the heat of the Sun's rays'* was read at the American Association for the Advancement of Science (AAAS), in 1856 in Albany, New York. Not by herself, as women were not allowed to speak at AAAS meetings, although they could be members. It was the first secretary of the Smithsonian Institution, Joseph Henry, that presented her paper instead, and probably failed to appreciate its key importance.

Her technique involved measuring the temperature of a glass cylinder of about 10 cm in diameter and 75 cm in length. The thermometers were placed in the cylinder which was then filled with different gases and put into shaded and sunny conditions. She makes three key points in the paper: i) *'the action* [the temperature change] *increases with the density of the air and is diminished as it becomes more rarified'*, ii) *'the action of the Sun's rays was greater in moist than in dry air'*, and importantly, iii) *'the highest effect of the Sun's*

---

[12] Foote, E., 1856, and see Jackson, R., 2019 for a more detailed analysis of the question of priority. Interestingly, Robert Jackson is also the biographer of John Tyndall. Jackson, R., 2018. Since the publication of Foote was only discovered as recently as 1992, and became more well known around 2016, he mentions the issue in two sentences on page 158 and on the last page of the book, p 455. We follow mostly the 2019 analysis of Jackson here.

rays I have found to be in carbonic acid gas'. The emphasis on the Sun's rays is important here and may be the reason why her findings did not get the attention they deserve. Her experiment did not distinguish, and was fundamentally unable to do so, between the effects of the short-wave and long-wave radiation from the Sun (the visible light and the infrared). This was one of the reasons why John Tyndall's experiments were more appropriate for determining the greenhouse effect. She did note, however, the possible impact of her findings for climate: '*an atmosphere of that gas* [carbonic acid] *would give to our earth a high temperature; and if as some suppose, at one period of its history the air had mixed with it a larger proportion than at present, an increased temperature from its own action as well from increased weight must have necessarily resulted*'. That is where Eunice Foote left it. There are no traces of additional experiments or further papers. However, her paper did get mentioned in various national and international newspapers just after it was published, both in the US and UK.[13]

John Tyndall apparently was unaware of the publication by Eunice Foote, although this is still a matter of some debate. On 21 September 1861 he managed to get his experimental set-up to produce stable results. Two months later he had determined the absorption of radiant heat (infrared radiation) by water vapour and compared it to that by dry air.[14] He was almost immediately aware of the importance of this result, as he wrote in the 1861 paper: '*But this aqueous vapour, which exercises such a destructive action on the obscure rays, is comparatively transparent to the rays of light. Hence the differential action, as regards the heat coming from the sun to the earth, and that radiated from the earth into space, is vastly augmented by the*

[13] Reed, E. W., 1992
[14] Jackson, R., 2018

aqueous vapour of the atmosphere. ... Now if, as the above experiments indicate, the chief influence be exercised by the aqueous vapour, every variation of this constituent must produce a change of climate. Similar remarks would apply to the carbonic acid diffused through the air; while an almost inappreciable admixture of any of the hydrocarbon vapours would produce great effects on the terrestrial rays and produce corresponding changes of climate.[15] He concluded that water vapour would greatly impact climate at geological timescales. He also emphasized the enormous impact of these colourless gases on the absorption of radiant heat and that their impact in small quantities was linearly dependent on the density of the gas. In interpreting his results, he noted that the German Kirchhoff[16] had shown that a molecule absorbs only waves that are synchronous with the resonant periods of vibration of that particular molecule. Referring to Dalton (see Chapter 3), he claimed that radiation and absorption can be reduced to the simplest mechanical principles. In this context it is worth noting that Kirchhoff's law states that emission and absorption are two sides of the same coin. This implies that when infrared radiation is absorbed, say by carbon dioxide, at a particular wavelength, it is emitted at the same wavelength. Apparently, Tyndall was on to establishing this attribute of radiation himself, claiming that only time prevented him from fully establishing the principle.[17] The explanation of this phenomenon would have to wait for Max Planck in 1900, but this dual role of absorption and emittance is at the heart of the greenhouse effect (see also Figure 2.3).

[15] Tyndall, J., 1861, see Jackson, R., 2018.
[16] Kirchhoff, G., 1860
[17] Jackson, R., 2018

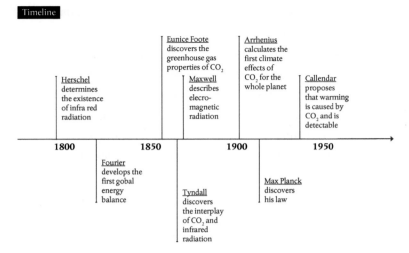

Figure 2.3 Timeline of the main events described in this chapter.

Let us go a little deeper into the physics of radiation. Kirchhoff, the man from the law that states that emission is equal to the reverse of absorption, developed the fundamental concept of a (perfect) black body. Such a body would absorb all radiation, and thus be perfectly black. However, given a certain temperature a black body would also emit radiation, and—if the temperature was high enough—emit radiation in the visible parts of our spectrum. Kirchhoff imagined this body to be a simple hollow container with one small, tiny hole in one of its walls. His key insight was that the intensity of the radiation inside his perfect black body depended only on the temperature and not on the material properties, such as shape or whether it was made from iron or wood. Given a certain temperature, the radiation inside the container would always be the same wavelength, and, importantly, also emitted at that wavelength through the single tiny hole.

How the radiation changes with temperature, can be observed by simply putting an iron poker into a fire. It would first be black, then start to glow a little red and then when the fire is very hot it may go towards the bluish-white part of the spectrum. What exactly the precise mathematical relation was between temperature and emitted radiation, however, was unknown. This became known as the 'black body problem': determining the spectral energy distribution of black body radiation, the amount of energy at each wavelength from the infrared to the ultraviolet, at a given temperature.[18] Understanding this became not only a scientific priority, but was also one with important industrial applications, particularly in the developing electric-light industry. If you developed a light bulb, you would for instance like to 'waste' as little energy as possible outside the visible range.

At the newly founded Imperial Institute of Physics and Technology, in Charlottenburg, near Berlin, these aspects were at the forefront of investigations. Wilhelm Wien, who worked there, discovered a simple relationship between the amount of energy contained in a particular wavelength and the temperature. The relation is deceptively simple: the wavelength at which the maximum amount of radiation is emitted always yields, when multiplied by the absolute temperature of the black body, a constant. The work was published in 1896[19] and is still considered an important breakthrough. Doubling the temperature will shift the corresponding peak wavelength by a factor of 0.5. Using as an example the temperature of the Sun's outer rim, at 6000 K, Wien's law predicts the peak wavelength to be at 0.5 μm (1 μm is 10^{-6} m), which is roughly the colour green, straight in the middle of the radiation

[18] Kumar, M., 2008
[19] Wien, W., 1896

that the human eye can perceive. And now the surprise: taking not the Sun's temperature but the temperature of the Earth at 288 K (note that we again use absolute temperatures here, to convert into °C, you only need to subtract 273.15), we obtain a value of 10 μm, which is now in the middle of the infrared—the invisible rays of Herschel! This spectral shift comes straight from Wien's law and is fundamental to understanding the impact carbon dioxide has on the climate. But more on that later.

First, we need some further understanding of how much energy is contained in these wavelengths. Calculating the wavelength of the peak emission is not the same as calculating the energy for the whole spectrum. For this we need the work of Max Planck. He had become interested in the black body problem and had started to tackle it using Maxwell's electromagnetic theory. Wien had established an early version of this so-called distribution law, but this ran into trouble at longer wavelengths, where it started to disagree with the experimental findings. These measurements were hugely complicated and of course not without errors. That caused something of a problem, as at these longer wavelengths the energies appeared to follow the Rayleigh–Jeans law. Planck's genius was that he combined them into one formula that contained only two constants. While this was, in a way, initially just a smart interpolation, it contained no real physical meaning yet. That was to come when Planck realized that he could use the concept of entropy and probability as expressed by Ludwig von Boltzmann.[20] His resulting formula contained it all: the law called Stefan–Boltzmann that related the total radiative energy from a black body to a product of a constant σ and the fourth power of temperature (σT^4), (this was discovered experimentally by Stefan, and theoretically

[20] Born, M., 1948

justified by Boltzmann in 1886), Wien's displacement law, and the two laws that expressed the energy density at high and low wavelengths from Rayleigh and Wien. They all could be derived or were encompassed by Plank's new formula. The two constants are now known as Boltzmann's constant and Planck's constant. The formula was a staggering success for Planck, something he himself appeared to be quite aware of. Normally known as rather a shy man, there is a later note from his son, saying that during a walk in the Grunewald (green forest) near Berlin, his father told him *Today I have made a discovery as important as that of Newton.*[21]

Planck's constant, with dimensions of energy multiplied by time, was originally called by him the 'elementary quantum of action'. It was to take on a phenomenal role in physics when expressed as a quantum of energy, because it implied that energy, E, could only be transmitted in small discrete chunks, the multiples of the product of their frequency, v and Planck's constant, h, ($E = hv$) and not as a continuous smooth progression. Quantum physics and quantum mechanics started with Planck, although he himself appeared for a long time to reject the fundamental idea of the quantum. Albert Einstein would soon fully appreciate the consequences of Planck's work. Max Planck was awarded the 1918 Nobel Prize for Physics *'in recognition of the services he rendered to the advancement of Physics by his discovery of energy quanta'.*[22] Exactly one century after William Herschel had first identified infrared radiation, Max Planck had elucidated the physics of radiation across the full spectrum.

Let us take stock. Maxwell identified radiation as an electromagnetic wave. Planck, using Maxwell's and Boltzmann's insights,

---

[21] Born, M., 1948
[22] Max Planck—Facts. NobelPrize.org, 2020

understood how radiation was absorbed and emitted by discrete minuscule quanta. We can calculate the energy content or density for individual wavelengths, using Planck's formula, and know that the Earth emits infrared (long-wave) radiation, and the Sun short-wave, visible radiation. The difference between Eunice Foote and Tyndall is also clear; Eunice Foote had measured the impact of the full spectrum from the Sun (with a spectral maximum in the visible part of radiation), while John Tyndall, using his cube of boiling water, used infrared radiation only. In short, we understand the basic physics of radiation, and know there are some gases, in particular water vapour and carbon dioxide that can absorb infrared radiation and importantly also radiate it back at the same wavelength. It is time to link all this information to climate and humans. This task was taken on by the Swede Svante Arrhenius.

Svante Arrhenius received the Nobel Prize in Chemistry in 1903 for work begun in his doctoral thesis of 1884.[23] This contained a ground-breaking theory (some call it the beginning of physical chemistry) on the dissociation of substances into electrolytes or ions. It appeared he had difficulty finding employment after getting involved in controversies with professors who held a rather low opinion of his PhD work. When, after some travels and periods working abroad, he returned to Sweden in 1891 as a teacher of physics in Stockholm, he withdrew from the field of electrolytes and solutions and became interested in the field of 'cosmic physics'. This is best described as a sort of Earth System Science *avant la lettre*, linked among others, to the ideas of the great German explorer and scientist Alexander von Humboldt that aimed to provide a more integrated picture of terrestrial, atmospheric, and

---

[23] Crawford, E., 1997

cosmic processes. Discussions within the scientific community in Stockholm put him onto the track of solving one of the great mysteries of the time: what caused the glacial and interglacial cycles geologists were starting to discover (Chapter 7). In discussions at the Physical Society in Stockholm he became interested in the subject, so he set out to see if small variations in the amount of carbon dioxide (he consistently refers to it as carbonic acid) could be responsible for variations in temperature of the order of 5–10 °C between glacial (cold periods with large ice sheets) and interglacial (warmer) periods.

Arrhenius thus set out to address the question: *'Is the mean temperature of the ground in any way influenced by the presence of heat-absorbing gases in the atmosphere?'*[24] In discussions about climate change, it is often said that by around 1900 the back of an envelope calculations of Arrhenius had already shown the possible impacts of burning fossil fuels on climate. One thing we can be sure about is that these people have not read Arrhenius's 1896 paper! Arrhenius himself referred to his manifold calculations as tedious, and there is a suggestion that he undertook them as a sort of therapy against the loneliness and melancholy resulting from divorcing his first wife. He started the work in 1895. To be able to perform his calculations he needed more than the results of John Tyndall's experiments: Arrhenius needed to calculate the impacts of the whole atmospheric column of air on the absorption of CO_2. Here the work of American Samuel Langley, who tried to determine the Moon's temperature measuring the outgoing infrared radiation, came in handy. Langley had developed a device, called a bolometer, with which he measured the infrared spectrum that arrived at the Earth from the Moon. Arrhenius made clever use of these

---

[24] Arrhenius, S., 1896

observations by analysing the impact of varying concentrations. As the elevation of the Moon declined, the radiation from the Moon would travel a longer path through the Earth's atmosphere, containing more absorbing gas. This allowed him to calculate the absorption. Langley calculated these for 21 'groups of rays', what we now would call spectral bands. Because the temperature of the Moon is not too different from that of the Earth, Arrhenius assumed that the use of Langley's data would be sufficient for his purpose.

Now it is time to go a little deeper into the core of it all and introduce some of the basic physics of the absorption of infrared radiation by greenhouse gases. There are several important issues at stake here. The first relates to the structure of the absorbing molecule that allows only particular vibrations to occur when a discrete packet of radiation is absorbed (there is only a limited amount of bending and stretching such a linear molecule can achieve). This creates specific absorption lines that represent those wavelengths at which the CO_2 molecule can absorb (and emit!) the radiation. These lines (more appropriately called bands) are located at 1.6, 2.0, 2.7, and 4.3 μm in the part of the spectrum towards the visible, called the near infrared (NIR). Towards the infrared (IR) they are at 9.4, 10.4, and 15 μm. Arrhenius undertook the difficult task of extracting this precise information from Langley's measurements. The second complication arises from the interaction of the absorption of water vapour and CO_2 in the same wavelengths. Water vapour also absorbs strongly in some of the bands in which CO_2 absorbs, so it is difficult to separate the two effects. Thirdly there is an effect known as pressure broadening that broadens the absorption bands when the pressure of the gas increases. Figure 2.4 shows how well, or how badly, Arrhenius did.

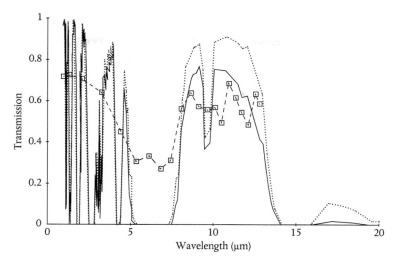

Figure 2.4 Spectral transmission by both H_2O and CO_2 as calculated by Arrhenius from Langley's observations (squares and broken line) and those calculated by a modern radiation code that calculates transmissivity for very small spectral intervals (the solid and dashed lines represent two assumptions about the water vapour content of the atmosphere, a high and low amount).
Redrawn after Dufresne, J.-L., 2009. See also Archer, D. & Pierrehumbert, R., 2011

The figure shows the transmission through the atmosphere, like the measurements of Langley that Arrhenius used, with a value 1 representing a completely transparent atmosphere (i.e., no absorption). The squares are the estimates that Arrhenius deduced from Langley's measurements. The continuous lines show calculations with modern radiation codes that are based on detailed spectral information and calculate the transmission for very small wavelength intervals. What the comparison shows is that Arrhenius generally did quite well at getting the peaks right, but calculated absorption in the 6 and 7 µm bands that he wrongly attributed to CO_2 (this absorption is large due to water vapour). He also completely missed the absorption of CO_2 in the 15 µm band because Langley's observations did not extend to that wavelength. However, the absorption in this region is very important for the radiation balance. So, he may not have been correct in all the details, but his final results, when corrected for these errors, are remarkably close to modern estimates. It appears he did quite well, given the limited data that he had available.

Having estimated the absorption of CO_2 and H_2O by the atmosphere, Arrhenius now set out to calculate the effects of changing the concentration of these gases. The way he approached this was to divide up the world into 10° latitude bands. For each band he calculated the effect. He assumed that the atmosphere had one single uniform temperature. Making this assumption allowed him to express the balance between incoming solar radiation and outgoing long-wave (infrared) radiation using the law of Stefan–Boltzmann that, as we have seen, expressed the radiation intensity as a function of the absolute temperature. Putting the equations together he then arrived at an equation that gave him the temperature difference as a function of the amount of solar radiation reflected (expressed as (1–albedo), where the albedo is the reflection coefficient), the incoming solar radiation and the absorption coefficients for solar and for infrared radiation. He incorporated some rather crude effects of clouds and was able to consider the effects of water vapour by first calculating the temperature change to CO_2 only, and then, assuming the relative humidity remained unchanged, the effect of water vapour. In fact, in performing these calculations—an estimated ten to hundred thousand by hand (!)—he established the basic way current climate models calculate the effects of increasing CO_2 level. This consists of establishing an energy budget equation and estimating radiation absorption, per grid square, rather than by latitude band (see Chapter 10). He did not repeat Fourier's mistake of assuming a significant contribution from intrastellar space, as he set that particular temperature to zero degrees Kelvin.

His calculations averaged over the Earth came to -3.2, 3.4, 5.5, 7.2, 8.4 °C if the amount of CO_2 absorption was 0.67, 1.5, 2.0, 2.5, and 3.0 times the actual value. An almost 30% reduction would then result in global cooling by 3.2 °C, while an increase would lead to progressively higher values of warming. He was careful enough to realize that only the values between 0.67 and 1.5 had been observed, or derived from Langley's values, and that the higher values were extrapolations. It has been said that these values, say for a doubling of the CO_2 concentration, are eerily close to what today's complex computer models predict. But he was lucky in that errors in his absorption estimates compensated and that the rest of his calculations also made assumptions that kept his estimates within reasonable bounds. He had, for instance, not considered the dynamics of the atmosphere at all. Climate models of course also consider exchange of heat, momentum, and water vapour and trace gases between the grid squares (Chapter 10), something Arrhenius was unable to do. Arrhenius performed his calculations initially with the aim of providing evidence that small changes in the amount of CO_2 in the atmosphere could trigger an ice age. He had precisely achieved that and illustrated this by ending his 1896 paper with a summary of a paper by his colleague Högbom, who was an expert on carbon cycling. He ends his paper rather modestly by saying that '*I trust that after what has been said the theory preposed in the foregoing pages will prove useful in explaining some points in geological climatology which have hitherto proved most difficult to interpret.*'[25] It was going to be much more than useful.

His 1896 paper was very much geared towards investigating the possibility that small changes in the CO_2 concentration of the

[25] Arrhenius, S., 1896

atmosphere could contribute to shifting the Earth into or out of an ice age. He was, however, aware of the other competing theories in this field, as his inclusion of the work of Högbom illustrates. But what did he say about the possibility that fossil fuel burning would possibly make the Earth warmer? Not much in his 1896 paper. While he calculates the warming for doubling and trebling of the CO_2 content he did not directly link this to fossil fuel burning. It was in fact Högbom who estimated the contribution of fossil fuel burning to the CO_2 load in the atmosphere.[26] However, his estimate was still so low at the time that he thought it was balanced by increased removal of CO_2 by erosion (Chapter 4). Arrhenius did go on, however, to calculate how long it would take with those estimated low rates to double the concentration in the atmosphere. Considering the role of the ocean in buffering (Chapter 9), he estimated that it would take 3000 years, to achieve this and the concurrent warming with it. In his 1908 book, *'Worlds in the making'*[27] he slightly revised the numbers: *'Although the sea, by absorbing carbonic acid, acts as a regulator of huge capacity, which takes about five-sixths of the produced carbonic acid, we yet recognize that the slight percentage of carbonic acid in the atmosphere may by the advances of industry be changed to a noticeable degree in the course of a few centuries'.* We are now talking about centuries rather than millennia, and Arrhenius—probably based on Högbom's thinking—suggests that for each 6 molecules of emitted CO_2, 5 end up in the ocean, with only 1 in the atmosphere (this is currently closer to 1 in 4, see Chapter 9). He continues in the book with this interesting statement: *'that would imply that there is no real stability in the percentage of carbon dioxide in the air, which is probably subject to considerable fluctuations in the course of time'.* CO_2 concentrations according to

[26] Crawford, E., 1997
[27] Arrhenius, S., 1908

Arrhenius have fluctuated in the past and will also in the future, causing changes in the global temperature.[28] In fact this is quite a statement given that he had no direct evidence of these fluctuations (see Chapter 7). While he was wrong in his estimates of the effect of fossil fuel burning on climate, he did the first proper calculations of the effect of increasing or decreasing levels of CO_2 in the atmosphere on the Earth's climate. Arguably it was a bit fortuitous that his numbers came close to current estimates of the effect, but, clearly, he is one of the founding fathers of climate science.

Not that everyone agreed with him. His theory that CO_2 caused ice age variability had to compete with others (Chapter 7) that were focused on orbital forcing (changes in the way the Earth moves around the Sun), solar variability, volcanic eruptions and dust, and several others.[29] Also, his extrapolation towards the impact of fossil fuel was not immediately picked up. He did receive support for his geological perspective in an extensive review by his Stockholm colleague Nils Ekholm who concluded towards the end of the review that 'the principal variations of the climate of the past, comprising a space of time of at least one hundred and perhaps one thousand million years, are probably due to long periodical, and perhaps accidental, variations of the quantity of carbonic acid in the atmosphere, whereas the insolation during all this time has been nearly constant....'[30] However, there is also secular change in the amount of carbon dioxide according to Ekholm. This is the increase in CO_2 due to the burning of fossil fuels. In accordance with Arrhenius, who based this estimate on the work of Högbom, he estimates that

---

[28] Arrhenius, S., 1908
[29] See Flemming, R., 1999. He lists in his table 9–1 12 such theories.
[30] Ekholm, N., 1901

this increased by 1/1000th each year. He notes that if this were to continue for thousands of years, a noticeable impact on the Earth's temperatures would be effected. He ends with a positive final statement that this gives man the opportunity to manage or regulate the climate and '... *and it will afford to Mankind hitherto unforeseen means of evolution*'.[31] Human-induced climate change as a positive issue.

It would take until 1938 before the possibility was raised that fossil fuel burning might not only increase the atmospheric CO_2 level, but also have a noticeable impact on the observed temperature of the Earth. The Englishman Guy Callendar was the first to make this relationship explicit. Callendar (Figure 2.5) was born in 1898, two years after Arrhenius had published his analysis.

Figure 2.5 Picture of Guy Callendar, taken in 1932
From Flemming, R., 1998

<hr />

[31] Ekholm, N., 1901

During the First World War he was enlisted as a scientist—he was medically unfit for active service—to work on X-ray technology. It was probably here that he became acquainted with Maxwell's electromagnetic theory on radiation. He worked under the supervision of his father at Imperial College, who held a professorship in physics and had been a very well-respected scientist and Fellow of the Royal Society since 1894. Guy Callendar initially worked on problems related to steam technology and helped his father with the Callendar Steam Tables, which were aimed at improving the efficiency of the steam engine and were much used at that time.[32] But his biggest legacy is without any doubt his work on the relationship between CO_2 concentration in the atmosphere and global warming. In his best known 1938 paper he spells out his goals quite clearly: *'Few of those familiar with the natural heat exchanges of the atmosphere, which go into the making of our climates and weather, would be prepared to admit that the activities of man could have any influence upon phenomena of so vast a scale. In the following paper I hope to show that such influence is not only possible, but is actually occurring at the present time'.*[33]

To achieve that he set out to do three things. First to prove that the CO_2 concentration had actually increased in the atmosphere, second to show that more CO_2 in the atmosphere changed the absorption and hence the climate, and third and last, to show that indeed temperatures had risen since humans started to use fossil fuels on a grand scale. Let us investigate each of these in some detail. To be able to show that the CO_2 levels had increased he needed accurate measurements that were representative of the natural atmosphere. Taking measurements close to a powerplant

---

[32] E.g. Heywood, H., 1945
[33] Callendar, G., 1938

was then, as now, not the best way to obtain a representative sample. Callendar used his knowledge of weather patterns to identify those observations that were likely to be free of contamination. In his 1938 paper he concluded that the best estimate of carbon dioxide in the free air of the North Atlantic at the beginning of the nineteenth century was 274 ppm (±5). This was later revised in his 1940 paper, using now more sets of observations, to 290.2 ppm. Selecting which measurement sites are reliable, which of those measurements were good (some of the measurements vary by more than 40 ppm between days, clearly indicating a measurement problem) was crucial in arriving at this number.[34] Using the 274 ppm as a starting point he then calculated how much CO_2 would remain in the atmosphere, taking into account possible ocean uptake. In doing this he assumed a linear increase in fossil fuel emissions, i.e. a constant annual addition of 4300 million tonnes CO_2 per year. His calculation showed that depending on the rate of uptake in the ocean, in 1936, 2000, 2100 the concentration would be 289, 314, and 346 ppm respectively. In taking the annual addition as a constant he followed Arrhenius before him, who had done the same. We now know of course that this is not completely right. While until roughly 1950 the increase was approximately linear, after 1950 the increase in fossil fuel emission follows an exponential curve. The impact of this mistake becomes clear in his calculation that includes a specific time reference: the current (2021) level of CO_2 in the atmosphere is 415 ppm, rather than the 320 ppm we would obtain if we followed Callendar, not considering changes in sinks etc. (Chapter 8).

---

[34] See From, E., and Keeling, C., 1986. According to these authors there is reason to suspect some of the data that was selected by Callendar as they failed to show the seasonality observed in modern data.

He then calculated the absorption using a different set of radiation measurements to those of Langley that were used by Arrhenius. One of the big improvements was that he calculated the absorption at the long wavelengths of 13–16 μm much more accurately. In doing so he was able to draw on some of the radiation measurements he encountered when he was working on steam engines. He then calculated the long-wave radiation coming from the atmosphere: *The method used to calculate the sky radiation from the absorption coefficients of water vapour and carbon dioxide is simple but laborious. It consists of dividing the air into horizontal layers of known mean temperature, water vapour, and carbon dioxide content, and summing the absorbing power of these different wave bands, in conjunction with the spectrum distribution of energy at the surface temperature.*[35] Unlike Arrhenius he divided up the atmosphere into several layers. He then used the change in CO_2 concentration to calculate the additional long-wave radiation the surface would receive. Figure 2.6 shows the results of his calculations. His calculations are accurate, but because he neglected the exponential increase in fossil fuel emissions, his predicted trend is somewhat off compared to reality. For instance, he calculated that towards the end of the twentieth century the increase in temperature would be 0.16 °C, rather than the 0.6 °C that is observed. However, if we make a correction for his neglect of other greenhouse gases and the exponential increase, his estimate would in fact come quite close, 0.52 °C as was shown in a recent analysis of his work.[36]

Callendar completed his analysis by looking at observed trends in temperature he obtained from the World Weather Records that were published by the Smithsonian Institution. Using a technique

[35] Callendar, G., 1938
[36] Anderson, T., Hawkins, E., & Jones, P., 2016

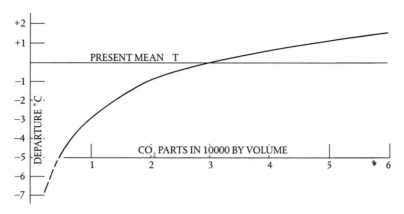

Figure 2.6 Callendar's estimate of the temperature change as a result of a change in CO_2 concentration in the atmosphere.
After Callendar, G., 1938

now common in climate science, to show the anomalies (deviations) compared to a longer-term mean, he was able to show that for the northern temperate zone, compared to the mean of 1901–1930, the last three decades of the nineteenth century were cooler, as were the first two decades of the twentieth century but less so, and that the periods from 1920–1929 and 1930–1934 were warmer by 0.2 and 0.5 °C (depending on which data he actually used, the latter numbers would be 0.16 and 0.4 °C, the signs of the series remaining the same). He was aware of the difficulty of showing that the Earth had warmed and warned towards the end of his paper that *The course of world temperatures during the next twenty years should afford valuable evidence as to the accuracy of the calculated effect of atmospheric carbon dioxide.*'[37] He could not have been more right.

[37] Callendar, G., 1938

CHAPTER 3

THE DISCOVERY OF THE CARBON DIOXIDE MOLECULE

T his chapter deals with the slow appearance of the carbon dioxide molecule on the world stage. Until the early seventeenth century, carbon
dioxide was not treated as a separate quantity—it simply did not exist in
the minds of people. Between the transition from alchemy to chemistry the
Flemish medical doctor Jean Baptista van Helmont discovered that something was released when charcoal was burned. He called that something
a gas, a gas sylvestris, a spirit from the woods, or wild spirit. The famous
developer of the air pump, and one of the founding fathers of modern science,
Robert Boyle realized that elements are substances that cannot be broken
down into simpler substances, and thus laid the basis for modern chemistry.
Around the middle of the nineteenth century the Scotsman Black discovered
that a white powder, magnesium alba, when heated would lose air, air he
called fixed air. He also experimented with sparrows, which he found would
die if exposed to fixed air. Some of the basic chemistry and effects of carbon
dioxide were being established, albeit in a rather animal-unfriendly way.

Like van Helmont, the German scientists Becher and Stahl realized that in combustion (burning) a substance was released. Becher had called this substance 'terra pinguis' (fatty earth), Stahl called it phlogiston. A chemical theory was born that explained many of the chemical transformations known at the time. By the mid-eighteenth century, most chemists accepted the phlogiston theory. But it contained some nasty, unexplained teething problems that triggered, the Frenchman Antoine Lavoisier amongst others, to perform some further investigations. He discovered that nothing like the mysterious phlogiston was released during combustion, but, in contrast, something was taken up that made the substance heavier. What was taken up turned out to be oxygen, the discovery of which should probably be credited to both Lavoisier and the Englishman Joseph Priestley. In 1774 Priestley discovered oxygen by heating mercury oxide. Lavoisier would follow a year later. Priestley later also experimented with fermentation and discovered further properties of 'fixed air', that was still not properly called carbon dioxide. Lavoisier's 'Elements of chemistry' also provided a further basis for the establishment of chemistry and name-giving of elements.

The Quaker John Dalton put everything on a more modern footing by realizing that the components of a substance would always appear as ratios of round numbers, like 2:1, or 3:1, never 1.39:2.77. He gave the first accurate description of carbonic acid, composed of one element of carbon and two of oxygen. It was then up to the Russian Dmitri Mendeleyev, in 1889, to put the final touch to our chemical understanding by developing the periodic table. The known elements of the world were now grouped according to their chemical properties. Carbon had an atomic weight (relative to hydrogen) of 12, oxygen of 16, so a molecule of carbon dioxide would weigh 44. The structure of carbon dioxide and indeed most of its properties, were now known, two centuries after van Helmont had first identified the gas.

It may sound strange now, given the recurrent appearance of carbon dioxide and climate change in the public debate and media, that carbon dioxide was not at all known as a separate entity until the middle of the seventeenth century. The honour of its initial discovery falls to the Flemish scientist and medical doctor Jean Baptista van Helmont. Van Helmont was quite a character, even considering the tumultuous years in which he lived. He was born in 1579, the year in which the Flemish people lost their hope of independence from the Spanish. In 1599 he graduated from the University of Louvain, only to leave the university a few years later in disgust at academic life. He travelled widely in Europe only to find *laziness, ignorance and deceit* all around him[1]. Unwilling to grow rich on what he perceived as the misery of his fellow men, he stopped practising medicine in 1605. Nevertheless, he continued to give unpaid advice to patients for the rest of his life. Married to the wealthy Margarita van Ranst in 1609, and becoming reasonably well off, he applied himself to a programme of natural philosophy that would allow him to 'unhinge' the works of nature and its instruments. In short, he turned to pyrotechnica, or chemistry.

He was caught up in disputes with the Spanish Inquisition and put under house arrest for most of the time between 1624 and 1642. The inquisitors objected to some of his early work in which he proposed to use experimental methods to test the validity of certain approaches. The case that brought him trouble was about the use of magnetic effects to cure wounds. The inquisition did not like the use of his experimental, proto-scientific methods, to prove or disprove hypotheses. He continued to work on his natural philosophy programme but published nothing during those years.

[1] Pagel, W., 2002.

Our knowledge of his work stems from two publications, one in Latin, the '*Ortus medicinae id est initi physicae inaudita*' (The origins of medicine, and the unheard-of principles of nature) and one in Dutch '*Dageraad ofte nieuwe opkomst der geneeskonst*', (Dawn of the new rise of medicine). Both were published after his death in 1644. The '*Ortus*' was first published by his son in 1648. The '*Dageraad*' followed in 1659 by a different route; allegedly in this case his daughter was involved in passing down the manuscript. His decision to write the latter in the vernacular Dutch rather than in the learned Latin characterizes his thinking: '*Dat immers alle inval van 't eerste gepeyns, gaende naer en tot woorden, zy altijdt eerst in de moeders taele*'.[2] Van Helmont was not a man to be attracted to, or bothered by, academic conventions (Figure 3.1).

In contrast to his predecessors, including Aristotle, who generally held that everything was composed of four elements, van Helmont only recognized two original elements, Water and Air, of which everything else was composed. Fire and Earth were in his view not original elements; Fire helped in separating elements, and Earth could be reduced to water. In one of his more famous experiments he let a willow tree grow in a container. He first weighed the soil (two hundred pounds of dried soil) and then planted a willow tree of 5 pounds. He let it grow for five years, weighed the tree again, made sure that there was still two hundred pounds of dried soil and determined that it had gained one hundred and sixty-four pounds of wood and bark. He attributed

---

[2] From the Dawn, Dageraad, freely translated: 'Because the first thoughts that translate into words are always in the mother tongue'. Van Helmont, J.B., 1660.

Figure 3.1 Jean Baptista van Helmont and his son Franciscus Mercurius who was responsible for publishing his father's work, the *Ortus Medicinae.*
From van Helmont, 1652

annes Baptista
Helmont

this weight gain to water: what else had been available to sustain the growth? He did not weigh the leaves and missed the small amount of nutrients that would have been taken up from the soil. This small amount was likely lost in the uncertainty of his measurements.[3]

He was, however, right about the important role of water, as we will see in Chapter 4 when we discuss photosynthesis. Also, and importantly, he was the first to coin the term gas. He identified the substance of a gas when it was released from the charcoal through burning: *'Gas becomes manifest when a solid body is made to relinguish its "vestments", the husk or shell that conceals it's essential (spiritual) centre.'*[4] When he put 62 pounds of charcoal in a closed vessel and burnt it, he was left with 1 pound of ashes, the remaining 61 pounds escaped in the form of smoke. Since this was neither water, nor

[3] Hershey, D.R., 1991
[4] Cited in Pagel, W., 2002.

air, he called this substance a *gas*. In this case van Helmont wrote about *gas sylvestris*, or a gas from the wood, sometimes known as *spiritus sylvester*, a wild spirit, also from the woods. Van Helmont identified about 15 different gases[5], while none of his predecessors had been able to distinguish between individual gases and only recognized air. He was thus the first to appreciate how gas was produced because of chemical processes. In an early treatise on bubbly water, he identified that certain minerals (iron bicarbonate) could be dissolved to produce gas. He also identified that this *gas sylvestris* was able to kill the flame of a candle, that it appeared in mines and that it originated when grapes were fermented. In short, he had identified what we now know as carbon dioxide.[6] In doing so he was one of the first to take the step from the alchemy of his predecessors to a more modern form of chemistry. Where previously only liquids and solids had existed, gases now also made an important contribution to the nature of matter. His use of balances for measuring weight gain also provided a major pointer for the further development of experimental science by better known greats such as Joseph Black and Robert Boyle. The latter not only appeared familiar with the work of his Flemish colleague but apparently also appreciated it: *'For I finde that Helmont ... an Author more considerable for his Experiments than many Learned men are pleas'd to think him...'*[7]

One of the problems that faced van Helmont in further investigating the properties of his gas sylvestris was the lack of an airtight instrument that would be able to collect and hold the gas. It was Robert Boyle, who developed the 'air-pump' that

---

[5] Partington, J., 1936
[6] Partington, J., 1936
[7] Boyle, R., 1661

would allow his. In 1643, the Italian Torricelli had produced the world's first barometer by letting the pressure of air outside come into equilibrium with a column of mercury. He discovered that the column would always rise, albeit with small daily fluctuations, to 76 cm, what we now know is the pressure of ambient air. The pressure of air was spectacularly demonstrated by the German Otto von Guericke in 1654, in front of a large audience, including the emperor Ferdinand III. Von Guericke had made two well-fitting copper hemispheres of 50 cm diameter each. These were connected to a pump that extracted the air inside. He then put eight harnessed horses on each side, but these failed to separate the hemispheres. In a final dramatic movement that would not have been misplaced in a theatre, he then let the air back into the hemispheres and they both fell apart to a stunned audience.[8]

Determining how carbon dioxide is formed and what chemistry, rather than alchemy, is involved, brings us to the heart of what historians and philosophers of science have called the scientific revolution.[9] In the case of carbon dioxide, the issue is to determine what exactly causes carbon dioxide to form and here we need to make an excursion into early combustion theories involving phlogiston[10] and the famous debates between Joseph Priestley and Lavoisier. Robert Boyle and his colleague Thomas Hooke would play an important role[11] in the development of scientific methodology they used to gain insight into how air and gases behaved.

---

[8] Strathern, P., 2001
[9] For the classic treatment see Kuhn, T., 1970., for a more recent, excellent overview Wootton, D., 2015.
[10] See for instance Conant, J., 1950, and Strathern, P., 2001 for overviews.
[11] See the celebrated, but not uncontroversial, account by Shapin, S., and Schaffer, S. 2011.

Robert Boyle in Oxford was well aware of the Guericke experiments and his pump, as he and his collaborator Robert Hooke set out to develop an even better pump attached to a vacuum flask. This air-pump was a very sophisticated device that allowed Boyle to do experiments with reduced pressure and about the effects of the absence of air on animals. Putting a variety of animals, such as ducks, frogs, adders, and even a small kitten, inside the machine, he discovered that most animals died within a short period of time. He further concluded that air was inhaled and exhaled from the lungs. Air was no mystical quantity as envisaged by Aristotle; it was a substance with specific properties of its own. Boyle concluded that elasticity was one of these; when compressed, air appeared to behave like a spring, it bounced back.

In his 1660 publication 'New Experiments Physico-Mechanicall, Touching the Spring of the Air, and Its Effects', Boyle described 43 experiments with the air-pump (Figure 3.2) that not only redefined the perception of air and vacuum, but also the way in which science, in particular experiments, could be used to exclude possibilities or disprove previous hypotheses, something that had brought van Helmont into trouble with the Inquisition. Shapin and Schaffer claim—a claim that is not undisputed among historians and sociologists of science—that Boyle was the first to establish a process designed to produce facts ('matters of fact'). Yet, he did not just rely on what they call the 'material technology', i.e. the technical construction of the air-pump, he also established not only a 'literal technology' by which he could convey his results to people who had not been witness to the original experiments, but, importantly, he also established a 'social technology' that 'incorporated conventions experimental philosophers should use in dealing with each other and knowledge claims'. All these so-called

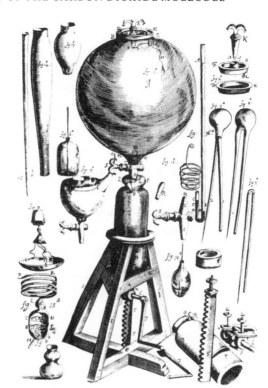

Figure 3.2 Drawing of Robert Boyle's first air-pump. Note the glass globe and the pump underneath it.
From Boyle, R., 1660

technologies produce knowledge. This way of producing knowledge underpinned the establishment of the Royal Society and one could argue presented a fundamental shift in the way science was done—hence the term 'scientific revolution'. This was a far cry from van Helmont's almost private experimentation at his manor house in Vilvoorde, the results of which were only published after his death. Science was becoming institutionalized.

Robert Boyle is sometimes called the father of chemistry. This is primarily because of his 1661 publication, the 'Sceptical Chymist'. Note that the 'al' from alchemy has now been dropped.

The *Sceptical* Chymist often reads like a sophisticated discussion with van Helmont. We have already noted how much appreciation Boyle had for van Helmont (his name appears no less than 57 times), even though he often disagrees. In the *Sceptical Chymist* he takes apart the theory of the four elements, or two as in van Helmont's view. '*... I now mean by Elements, ..., certain Primitive and Simple, or perfectly unmingled bodies; which not being made of any other bodies, or of one another, are the Ingredients of which all those call'd perfectly mixt Bodies are immediately compounded, and into which they are ultimately resolved....*'[12] Elements are substances that cannot be broken down into simpler substances. It is often noted how close this is to our modern way of looking at molecules, but we have to be careful not to put too much into the words of Boyle. While he made a major step in realizing that compounds could be made from several elements and could also be retrieved in the reverse process, he basically had no idea what an element was or what it consisted of.

It would take almost another 100 years to arrive at the next step in determining the nature of carbon dioxide. Born in 1728 in Bordeaux, Joseph Black was in his later years a firm constituent of the Scottish Enlightenment that included illustrious people like Adam Smith, the writer of '*An inquiry into the nature and causes of the Wealth of Nations*', and his friend, the philosopher David Hume.[13] Black wrote his PhD thesis in 1754[14] on the chemical behaviour

[12] Boyle, R., 1661
[13] Rasmussen, D., 2017
[14] Guerlac, H., 1957, notes the difference between the Thesis (in Latin) and the text that is mostly cited and is in English: Experiments upon Magnesia Alba, Quicklime, and Some other Alcaline Substances. His theory can also be found in Lectures on the elements of chemistry, although this was derived from his manuscripts by John Robinson, and published in 1806 and may contain some interpretations of its editor, see Joseph Black's 'Lectures on the Elements of Chemistry', Partington, J.R., 1960.

of common alkalis (carbonates) and the alkaline earths, lime and magnesium. According to the historian of science Guerlac 'The Experiments upon Magnesia Alba' (magnesium alba is now known as magnesium carbonate) was quickly recognized as a 'brilliant' model, and even the first successful model of quantitative chemical investigation. Guerlac thought The Experiments were on par with Newton's Optiks as far as experimental science went.

Black was able to show by careful experimentation that the changes produced in magnesia alba by heating were associated with the loss of an elastic, aeriform constituent, a 'fixed air'. In his lectures that were published after his death in 1799, he explained why he called it fixed air, which is: 'any elastic matter, capable of entering into the composition of bodies, and of being condensed in them to a solid concrete state, by its chemical attraction for some of their constituent parts'. By then he had realized that the name fixed air was perhaps not very carefully chosen, as it concerned matter in an 'elastic form' so it was later changed to gas. He did however not cite van Helmont as the first person to coin the term gas. Perhaps more importantly, by carefully weighing the solids and the resulting gases, he was able to quantitatively show that the loss of the gas plus the remaining solid weighed as much as the original material.

At this point we need to introduce some basic chemistry. The first reaction that Black investigated was that of combining magnesium carbonate with sulfuric acid (magnesia alba and vitriolic acid). In today's chemical shorthand this reads as $MgCO_3 + H_2SO_4 \leftrightarrow MgSO_4 + H_2O + CO_2$, which states that when one adds one molecule of magnesium carbonate to one molecule of sulfuric acid, one obtains one molecule of magnesium sulfate, one of water, and one of carbon dioxide. Joseph Black was not familiar with these elements in the way we have just noted them down, so what we are doing here is something known as 'Whig history', describing past events with the knowledge of today. However, it helps to understand Black's genius, that without knowing the elements as we know them today, he was able to determine how CO_2 would form. He also realized that the reaction could also be reversed, as indicated by our double arrow. When he heated the magnesium carbonate he also obtained fixed air: $MgCO_3$ + (heat) $\rightarrow$ $MgO + CO_2$. This, he realized, was similar to the reaction that occurred when he heated chalk (calcium carbonate) to obtain quick lime and carbon dioxide: $CaCO_3$ + (heat) $\rightarrow$ $CaO + CO_2$.

Black discovered some key properties of fixed air, similar to those that van Helmont had discovered when he burnt his charcoal and produced the *gas sylvestris*. Black used sparrows to investigate the properties of the air. In an experiment in 1557 that probably would not pass today's rules for animal experimentation: '*I had discovered that this particular kind of air, attracted by alkaline substances, is deadly to all animals that breath it by the mouth and nostrils together; but that if the nostrils were kept shut, I was led to think that it might be breathed with safety. I found, for example, that when sparrows died in it in ten or eleven seconds, they would live in it for three or four minutes when the nostrils were shut by melted suet*'. He realized that by breathing common air, part of it was converted into fixed air and that if blown into limewater, lime would precipitate (limewater is an aqueous solution of calcium hydroxide, which, with carbon dioxide forms calcium carbonate (lime)). It would appear that he had identified the main chemistry and some of the properties of carbon dioxide but let us not forget that he was not aware of the existence of the elements as we write them in our shorthand chemistry of today. Thus, much about the nature of carbon dioxide remained to be discovered, above all its precise relation to combustion and respiration.

Combustion had always been something of a mystery. Greek natural philosophers held that all inflammable substances contained the element fire, which when the substance was exposed, to say, heat, was released. With van Helmont, fire had already been taken out of the key elements. It was up to two Germans to develop a new theory of combustion, Becher and Stahl. The crucial insight here was most likely by Stahl who realized that the process of extracting iron from ore by smelting is the opposite of combustion. In the case of smelting, the ore absorbs something or a substance from the charcoal, used to heat the ore, to become metal. He also surmised that rusting was a similar process to combustion but taking place at a different speed. In combustion that substance was released. Becher had called this substance 'terra pinguis' (fatty earth); Stahl called it phlogiston. This was a remarkable step: the phlogiston theory appeared to explain a lot of the transformations that people knew. However, there were also some nasty unexplainable issues: when iron rusted it became heavier rather than lighter, so it appeared to take up something that added weight. When wood was burned this was not a problem, what was left as ashes was a material mostly devoid of phlogiston. The increase in weight in van Helmont's tree experiment could also be attributed to the uptake of phlogiston from the air. Nevertheless, the phlogiston theory was generally accepted around the mid-eighteenth century by most chemists, despite the problems that we now know were the basis of its final refutation.

The Frenchman Antoine Lavoisier (1743–1794) took the issue of the weight increase more seriously. In a note written on 1 November 1772 he states that: '*About eight days ago I discovered that sulfur in burning, far from losing weight, on the contrary, gains it; it is the same with phosphorus; the increase in weight arises from a prodigious quantity of air that is fixed during combustion and combines with the vapors'*. He added in the same note this may hold for all other substances that gain weight during combustion. Rather than that something such as the mysterious phlogiston was released during combustion, in fact, something was taken up. While this insight was crucial, he still had not yet discovered what that substance was.

Enter at this point in the story, the Englishman Joseph Priestley (1733–1804), who according to one of his early biographers[15] and the natural philosopher Thomas Hobbes qualifies as the 'experimentarian philosopher' *par excellence*. While his lasting fame rests on the discovery of oxygen, to which we will turn shortly, it was with Black's fixed air that he began his pneumatic chemistry experiments. From his first paper in the Philosophical Transactions of the Royal Society we quote: *'Fixed air is that which is expelled by heat from lime, and other calcareous substances, and, when deprived of which, they become quicklime. It is also contained in alkaline salts, and is generated in great quantities from fermenting vegetables; and being united with water, gives it the principal properties of Pyrmont-water. This kind of air is also well known to be fatal to animals.'* Pyrmont water is what today we would call fizzy, or sparkling, water and named after the small town of Bad Pyrmont in Lower Saxony, Northern Germany. Priestley would be awarded the prestigious Royal Society Copley medal in 1773 for his work on gases: *'on account of the many curious and useful Experiments contained in his observations on different kinds of Air'*, as the Laudatio mentioned.[16]

[15] Thorpe, T., 1906
[16] Johnson, S., 2009

He could do further experiments on fixed air because he lived near a public brewery, where he noted that a large layer of gas always formed on top of a fermenting liquid—this layer would grow to about 30 cm deep.[17] He noted that it must have been heavier than air because it did not mix. Priestley also noted that when a flame or burning chip of wood was placed in the layer of fixed air, it would quickly extinguish. He had identified one of the crucial aspects of carbon dioxide, namely that it dissolves in water. Again, in his own words: *'Considering the near affinity between Water and fixed air, I concluded that if a quantity of water was placed near the yeast of the fermenting liquor, it could not fail to imbibe that air, and thereby acquire the principal properties of Pyrmont, and other medicinal mineral waters. Accordingly, I found, that when the surface of the water was considerable, it always acquired the pleasant acidulous taste that Pyrmont water has.'* While producing in the space of two to three minutes a glass of *'exceedingly pleasant sparkling water'*, of course this process of dissolving CO_2 in water is one of the key mechanisms of the ocean carbon cycle. He also described that when iron is put into a water solution containing fixed air it readily dissolves. He furthermore realized that the substance may have the nature of a weak acid. This is the carbonic acid that forms when carbon dioxide dissolves in water.

In 1774, on the first of August to be precise, Priestley discovered oxygen, a discovery that made him justly famous. He heated mercury oxide by focusing solar radiation using a magnifying glass. In today's chemistry notation the reaction reads as $2HgO \leftrightarrow 2Hg + O_2$, which translates as two molecules of mercury oxide form two atoms of mercury combined with one oxygen molecule (composed of two oxygen atoms). Mercury oxide is a red powder.

[17] Priestley, J., 1772

When this was heated oxygen was released, although Priestley at the time did not recognize it as such. He would always, till his death, keep on referring to it as dephlogisticated air. Priestley noticed that when he heated the oxide, silver shining mercury globules started to appear and that a gas formed. When he analysed the properties of that gas, he found it was a superior kind of air that would not dissolve in water (i.e., unlike another similar substance he had recently produced, a gas we now know as nitric oxide (NO) and which he called nitrous air). A mouse put into the flask that contained the newly discovered gas also lived longer than usual, i.e. a full hour compared to otherwise something close to 15 minutes. We know that he mentioned his findings in a letter to John Pringle,[18] physician to the king, George III. They were, in a slightly modified form, published later in his 1776 memoir entitled, 'Experiments and observations on different kinds of air'. He explains in his preface why he chose to publish this book separately and not through the Royal Society: 'One reason for the present publication has been the favourable reception of those of my Observations on different kinds of air, which were published in the Philosophical Transactions for the year 1772, and the demand for them by persons who did not chuse, for the sake of those papers only, to purchase the whole volume in which they were contained. Another motive was the additions to my observations on this subject, in consequence of which my papers grew too large for such a publication as the Philosophical Transactions.'[19] He seemed well aware of the importance of his findings and the need to make them available to the wider public.

A few months after his initial discovery, in October 1774 he made a trip around Europe on which he met members of the

[18] Priestley, J., 1775
[19] Priestley, J., 1776

Académie des Sciences and the French scientist Antoine Lavoisier. Allegedly during a dinner with the latter, he mentioned his mercury oxide experiment: '*As I never make the least secret of any thing that I observe, I mentioned this experiment also, as well as those with the mercurius calcinatus, and the red precipitate, to all my philosophical acquaintance in Paris, and elsewhere, having no idea at that time, to what these remarkable facts would lead.*' Lavoisier then continued his own investigations with mercury oxide and in the spring of 1775, Lavoisier also made the discovery of oxygen, although he too was not yet fully aware of his breakthrough as he thought that gas he produced was common air, and not a specific new gas species. This 'mistake' was due to the methodology he used to distinguish oxygen from common air (a somewhat tricky process that used the oxidation of nitrous air (NO) to laughing gas (N_2O) and determining the loss in volume over water). Lavoisier submitted a paper to the Académie with the clear title: '*Sur la nature de principe qui se combine avec les Métaux pendant leur calcination, & qui en augmente le poids*' (On the nature of the substance which combines with metals during calcination and increases their weight). Priestley subsequently commented on the paper and Lavoisier then corrected the mistake in a 1777, May paper.

In another experiment he investigated the effect of a burning candle that was put on a float over water. He would eventually call the gas oxygen from the Greek oxy (acid) and gen (generator).[20] According to Strathern,[21] the death of the phlogiston theory was celebrated in a 'rational scientific ceremony' with

---

[20] See Strathern, P., 2001, also for the fact that in the end, the use of oxy (acid) turned out being somewhat of a misnomer as other acids exist without the presence of oxygen (e.g. hydrochloric acid).
[21] Strathern, P., 2001.

the wife of Lavoisier (his partner in crime) dressed up as an ancient Greek priestess ceremonially burning the works of Becher and Stahl (Figure 3.3). While generally, and correct in timeline, Priestley is credited with the discovery of a gas he called dephlogisticated air, Lavoisier must be credited with working out the details and realizing it was not dephlogisticated air but oxygen. However, the story of its discovery is not complete without mention of the unlucky Swedish scientist Karl Scheler who discovered several unknown elements, but also oxygen, which he called 'fire air'. He produced it in a similar way as Priestley and Lavoisier in 1772 from mercury oxide, but his publication only came out after severe delays in 1777, when Priestley and Lavoisier had already laid claim to the discovery.

Figure 3.3 Celebrated picture by J.-L. David, a double portrait of Antoine Laurent Lavoisier and his wife Marie Anne Pierrette Paulze.

From the Metropolitan Museum of Art, New York.

Some two centuries after van Helmont first described his *gas sylvestris*, several species of gas had now been discovered, and the fundamental process of oxidation had been established. But the name-giving of the different species of gas was all over the place. It was again Lavoisier who put order into this chaos by proposing a more consistent and logical system of naming in which compounds would be recognized by the conjunction of their element names. This truly revolutionized chemistry and made it into a full-blown science, based on experimentation and strict rules of deduction. It is difficult to underestimate the impact of Lavosier's *'Elements of chemistry'* (the original in French is named *Traité élémentaire de chimie'*) that was published two years later. Not everybody, however, shared that opinion, and in 1794 he was beheaded by the guillotine in the aftermath of the French Revolution by the Jacobins, the judge claiming that 'the republic has no need of a scientist'.[22] His marriage into the aristocracy, political allegiance, and his role as tax collector no doubt provided the stronger arguments for putting his head under the guillotine.

While Lavoisier's *'Traité élémentaire'* was undoubtedly a masterwork and instrumental in the further development of chemistry, we are still a long way off our current understanding of carbon dioxide and oxygen. Lavoisier identified the law of conservation of mass that solved the issue of metals gaining weight when oxidized. He provided a novel table of elements that finally did away with the classical four elements and, however incomplete and rudimentary, forms the basis of what we use today and put the phlogiston

---

[22] It is interesting to note that the French revolution played an important role, not only in the death of Lavoisier, but also in the life of Joseph Priestley, albeit for the opposite reason in the case of Priestley. His house in Birmingham, including his laboratory, was burned down by a mob in 1791 for his support of the revolution. He decided to emigrate to the US in 1794. See Thorpe, T., 1906.

Timeline

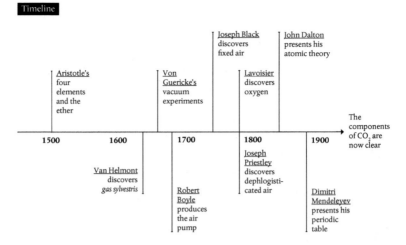

Figure 3.4 Timeline of the main events described in this chapter.

theory to bed. No minor details, but the system we use today, with names such as nitrous oxide (N_2O) and of course carbon oxide (CO_2), a compound composed of carbon (C) and two (di) oxygen atoms (O_2) was still some way off. It was a Quaker, John Dalton, who would provide that part of the puzzle (see also Figure 3.4).

Born in 1766, Dalton developed his theory of the atomic model around 1800. It was first published in 1801 and he presented it in 1803 and 1804 at the Royal Institution in London, where another famous chemist, Davy, was working. 'On the absorption of Gases by Water and other Liquids' was also read to a small audience in the rooms of the Literary and Philosophical Society of Manchester on 2 October 1803, and printed in the Manchester Memoirs in November 1805. Dalton thought that elements such as mentioned by Lavoisier were made of atoms, and importantly that one of their distinctive features was their weight. Individual atoms could not be destroyed nor created. They could, however, combine into compounds. Dalton's insight explained the Law of simple proportions that was developed by the French chemist Louis-Joseph Proust. This law stated that the compounds would always contain elements in simple ratios like 3:1, but not like 3.18:2.87. If you, as Dalton did, assumed that gases and other elements were composed of small, indivisible particles, this made perfect sense. Each element has a specific weight, so the weight of a compound would always be a combination of the weight of the constituent elements.

Dalton is also famous for Dalton's law which states that the total pressure of a combination of gases is equal to the sum of the partial pressures of each contributing gas. In a real atmosphere this means that nitrogen (80%) and oxygen (19%) are responsible for most of the pressure, with CO_2 contributing almost nothing. Dalton also wondered why CO_2 dissolves so well in water in the case of lime water, but that water under a normal atmosphere would hardly take up CO_2. It is worth quoting his reasoning in full from his 'A new system of chemical philosophy'.[23] In 1802 William Henry, the chemist and medical doctor, had published a paper in the Philosophical Transactions of the Royal Society that stated that the amount of gas absorbed in water increased in direct proportion to the pressure that such a gas exerted on the water. Dalton was at that time struggling to understand why lime water contained such large quantities of carbonic acid, while pure water contained hardly any. However, he made an important discovery: *'In pursuing the subject I found that the quantity of this acid taken up by water was greater or less in proportion to its greater or less density in the gaseous mixture, incumbent upon the surface, and therefore ceased to be*

[23] Dalton, J., 1842

surprised at water absorbing so insensible a portion from the atmosphere.' Dalton then goes on to generalize the observation and makes it a generally applicable law: '...... *that each gas in any mixture exercises a distinct pressure, which continues the same if the other gases are withdrawn'.* This is now known as Dalton's law.

What we read here is good example of the workings of the mind of an almost modern, creative, and solid scientist. Apparently, Dalton had done some experiments for which he had found an explanation, but only realized when Henry came up with a generalization, that his law of partial pressures would provide a very good explanation and he went on to do the additional experiments. Increasing partial pressure of CO_2 leads to increased oceanic uptake, as we will see later. Dalton's analysis added to Priestley's experiments providing further insight into this oceanic uptake process.

Dalton was in a way the first to provide a modern image of the CO_2 molecule. Figure 3.5 shows his system in graphical form with his explanation. Dalton gave the hydrogen atom, known as the lightest element, a weight of 1; this allowed him to relate the weight of the other known atoms to hydrogen. He knew that water had weight proportions of hydrogen to oxygen 1:7, this implied that the weight of oxygen would be 7. The interesting part here is that Dalton was wrong—he assumed water was HO, rather than H_2O (item 21 in Figure 3.5; the ratio of hydrogen to oxygen atoms in water is 2:1 rather than 1:1; hence oxygen should be 16, as we now know). Here, items 25 and 28 are of interest. Item 25 in his list is what we now know as CO, carbon monoxide; item 28 is CO_2. The image of item 28 is surprisingly modern, one atom of carbon bounded by an atom of oxygen on each side. For the first time in history, Dalton had now produced a system that: identified atoms

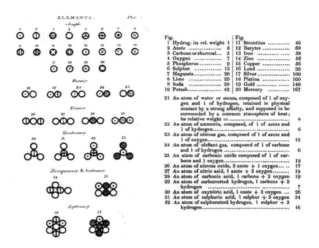

Figure 3.5 Table of elements showing the make-up of compounds out of elementary building blocks, the elements or atoms.
See Dalton, J., 1842.

as the singular building blocks of compounds that can neither be destroyed nor created and only be combined in whole numbers; and presented a (relative) mass for each, that allows 'precise' determination of the quantities of the elements/atoms needed to produce a compound. Dalton, being a Quaker, shunned prestige and public honours. Yet in honour of his memory, today, chemists and biologists all over the world refer to the Dalton as the unified atomic mass unit, which is defined as $1/12$ of the mass of a carbon atom and weighs approximately $1.66053886 \times 10^{-30}$ grams, a tiny, very tiny amount of mass.

Dalton had put chemistry on a solid basis, despite the inconsistencies that his system still contained. However, the question why these elements precisely formed compounds and how they stayed together was not yet resolved. This required some more work, and for now sidestepping the work of Berzelius, who identified that atoms may have different charges (electricity), Avogadro, who realized that any given volume of any gas contains the same number of molecules or atoms, and the German Meyer, who devised a new periodic system himself, but appeared too concerned about the anomalies in his system to publish it quickly, we arrive at the work of Dmitri Mendeleyev, the developer of the periodic table. Having struggled for some time to establish order in a seemingly chaotic multitude of elements with different masses, he apparently had a dream into which everything fell into place: *'I saw in a dream a table where all the elements fell into place as required. Awakening, I immediately wrote it down on a piece of paper.'*[24] His table is known as the periodic table, periodic because when elements were put into the order of their weight, their properties repeated themselves in a periodic manner. With hindsight, he was to use it to predict several new elements and the properties of their compounds. At the time, Mendeleyev also corrected the atomic weights of some already known elements. This was the ultimate system that combined insight into the mass of elements with their chemical behaviour and every high school student will remember it from their chemistry class.

Mendeleyev published his table in the *Principles of Chemistry* in 1889. By now carbon was allocated a mass of 12, and oxygen 16, as we still use today. One of the key innovations of the periodic table is the grouping of elements of different mass, but similar valence (the electrical charge). So, in the group of halogens, F, Cl, Br, and I, (fluorine, chlorine, bromine, and iodine) the valence is 1 (they appear always as negatively charged atoms of charge −1 in a compound molecule). Going up one line, we get to the oxygen and

---

[24] Quoted Strathern, P., 2001.

sulfur group that has a charge of −2, and one above, the nitrogen group −3. Then we have carbon, on a charge of 4, and the group with boron and aluminium (B, Al) with a charge of +3.

The structure and the composition of carbon dioxide was now known. It comprises two atoms of oxygen and one of carbon: CO_2. The structure of a carbon dioxide molecule is linear with the carbon atom sitting in the middle between two oxygen atoms in a straight line. This is not always the case in other molecules with three atoms; in water, for instance, the two hydrogen atoms combine with one oxygen atom making an angle of 105°. The distance between the oxygen and carbon atoms in CO_2 is 116.3 pm (1 picometre is 10^{-12} metre, or one trillionth of a metre). This is an average distance, because under the influence of infrared radiation the molecule can stretch (and shrink) in the longitudinal direction. In fact, this vibration is key to the greenhouse gas effect of CO_2 as we saw in Chapter 2.

CHAPTER 4

CARBON DIOXIDE AND THE ROCKS OF THE EARTH

W e come to realize that carbon comes in many forms that continuously cycle around on, and in the Earth. Antoine Lavoisier appeared to be the first to make the connection between the carbon found in chalks and marbles and the carbon dioxide in the air. But the real fundamental insights of matter cycling around on Earth came from James Hutton, the founder of geology. He realized on his Scottish farm that exposed rock was continually being eroded and that the eroded material would have to be transported into the seas, where adding layer after layer produced compaction and eventually new rock. These sedimentary rocks would in turn also be eroded: there was no beginning, no direction, and no end. This of course did not go down well with his colleagues who still believed that the Bible contained most of what there was to be known about geology. Hutton is also remembered for his theory of uniformatism—processes in the past are similar to those at present—and he linked geological forces that produced mountains to heat inside the Earth. He was wrong about the origin of that heat, though, but he did realize the importance of coal formation.

The director of the French Royal Porcelain Works in Sèvres, Jacques Ébelmen did establish the key elements of the carbon cycle. He also hinted at the existence of a relation to climate. The Frenchman Brongiard, before him, in the 1820s, had realized that the deep deposits of coal in the Devonian period would originally have been able to retrieve carbon dioxide out of the air—as plants currently do. Burying carbon (through geological processes) he realized would affect the atmospheric concentration, and vice versa also that of oxygen. Roughly at the same time, the existence of carbonate in sea water was established by Alexander Marcet, while Jean Baptiste Boussingault hung his instruments above volcanic exhaust to determine that they mostly contained carbon dioxide. Something thus linked the carbon dioxide in the air with the rocks from below. Ébelmen now realized that chemical reactions between the material that formed the rocks and carbon dioxide could take it out of the air and eventually move it into the ocean. The geological part of the carbon cycle was beginning to be established. Harold Urey would finally put this into the now classic Urey equations that describe how carbon dioxide can react with rock material (silicates) and thus impact the atmospheric concentration. Water is needed in this process of weathering and the combination of availability of water and carbon dioxide provides a key to what we call the geological thermostat. This is the long-term mechanism that could keep the Earth's temperature in balance: a warm planet with high CO_2 has high precipitation; however, with both liquid water and CO_2 around, the CO_2 is taken out, hence lowering the concentration and temperature until it becomes too cool and dry and CO_2 starts to rise again. CO_2 plays a pivotal role in the Earth's climate.

The Russian scientist Vladimir Vernadsky established that life was not only driven by geology, but also the other way around—it could act as a geological force itself, thus also shaping geology. He coined the term biosphere *for all the living matter on the planet. His friend Victor Goldschmidt finally put two and two together and linked the organic carbon cycle (which consists of producing organic biomass and respiring it) with the geological cycle and with the concentration of CO_2 in the atmosphere. His final estimates of the size of the Earth's carbon reservoir were eerily close to today's estimates. It was now firmly established: carbon continuously cycles around in different forms. Hutton was right after all: 'no vestige of a beginning, no prospect of an end'.*

In 1793, a year before his untimely death, Antoine Lavoisier first hinted at the existence of a geological carbon cycle, or at least at the connection between combustion, respiration, and the carbonates found in the Earth in the form of chalks and marbles: *'The carbon, accordingly, must be considered as a simple combustible body, or at least that chemistry has not succeeded yet to decompose, that forms fixed air or carbonic acid by combustion [...]. We conceive after that the immense quantity of carbon locked in the bowels of the earth since marbles, chalk or calcareous rocks contain around three-tenth, and sometimes even a third of their weight of fixed air, and that this latter is composed of carbon for 28/100th of its weight [...]. These consequences are far from all the ideas that we had until now; but they are not a less necessary consequence of the experiments that we have just reported. [...] We will not follow here the changes of forms underwent by carbon when it passes from the mineral kingdom, to the plant kingdom and the animal kingdom.'*[1] What the exact

---

[1] 'Le charbon, d'après cela, doit être considéré comme un principe combustible simple, ou au moins que la chimie n'est point encore parvenue a décomposer, qui forme par sa combustion l'air fixe ou acide carbonique [...] On conçoit, d'après cela, quelle immense quantité de charbon se trouve renfermée dans les entrailles de la terre puisque les marbres, les craies, les terres calcaires contiennent environ trois dixièmes et quelques fois même un tiers de leur poids d'air fixe, et que ce dernier est compose´ pour les 28/100 de son poids de charbon [...]. Ces conséquences s'éloignent beaucoup de toutes les idées qu'on avait eues jusqu' à présent; mais elles ne sont pas moins une suite nécessaire des expériences

relation was between the carbonic acid in the air, the actions of plants to produce oxygen and use it to provide energy, and the vast carbon stores in the deep Earth was the next enigma that needed to be solved.

For the beginnings of that search we need, once again, to go back to the Scottish Enlightenment, and what was called the Oyster dining club. Not only was Joseph Black, next to the economist Adam Smith (Chapter 3), one of the initiators of this illustrious discussion group, it also included another famous person we have not discussed so far: James Hutton. These three people, each in their turn can lay claim to have founded a discipline: Black modern chemistry, Smith political economy and Hutton modern geology. The name Oyster club presents somewhat of a conundrum as the name suggests some appreciation for culinary delights, but Black was a vegetarian, Hutton an abstainer, while Smith's only weakness appears to have been a craving for lump sugar.[2] With these three founding members, it is no surprise that the conversation in the club was mostly about scientific issues.

James Hutton had studied medicine in Edinburgh, his town of birth, after an initial failed attempt to become a solicitor (Figure 4.1). He continued his studies in Paris, Utrecht, and Leiden where he obtained his degree in 1749 with the thesis 'De sanguine

qu'on vient de rapporter. [...] Nous ne suivrons pas ici les changements de forme que prend le charbon en passant du règne minéral dans le règne végétal et dans le règne animal.' Lavoisier, A.L., 1793. Sur le charbon. In: Euvres de Lavoisier, Vol. 5, 1892. Cited in Galvez, M. & Gaillardet, J., 2012.
[2] Rae, J., 1895

Figure 4.1 Image of James Hutton sitting at his desk by Henry Raeburn. Probably painted in 1797, not long before Hutton's death.
From the National Portrait Gallery of Scotland

et circulatione in microcosmi'.[3] From 1750 to late 1767 Hutton worked at his ancestral farm, which became his property after his father's death. It was during this period that he must have developed the ideas that would start modern geology. Unfortunately, no surviving set of diaries allows us to trace his thinking in detail. We can surmise that working at his farm, he noticed that some parts of the Earth were visibly eroding, losing soil. He subsequently reasoned that if that process were to continue, the planet eventually would become inhabitable. There had to be some process that restored this! Hutton reasoned that the eroded material, the sediments, would end up in the sea, where they would be consolidated, hardened, and, importantly, uplifted again by sub terrestrial heat (heat under the surface of the Earth, the details of which he did not always specify). Once the material was thus returned to the surface of the continents the cycle of erosion could start again. This cycle of rock and eroded material is now called the geological cycle.

---

[3] 'The blood and the circulation of the microcosm'. See Repcheck, J., 2003, which provides a fairly recent biography of James Hutton. Repcheck suggests that the notion of self-sustaining cycles may have first appeared here, later to be applied to the Earth.

In 1785 Hutton was invited to present his work to the Royal Society of Edinburgh in two lectures.[4] That is to say, the first of these two lectures was given by his friend Joseph Black, because of severe illness; the second he presented himself. Since he was not a very gifted communicator himself, and his friend, in stark contrast, was, it may ultimately have been to his advantage that the first part was read by Joseph Black. The transcription of the lectures appeared in the first volume of the Royal Society of Edinburgh three years later and contained the famous lines expressing his views on the successions of erosion, consolidation and lifting: *'no vestige of a beginning, no prospect of an end'*. This is the key characteristic of a cycle where, wherever you start, you can jump on the wagon and continue the cycle until you are back at the point where you started, and then everything starts again, although as Hutton stated, there is in fact no real start and end in a cycle.

Hutton was up against some strong opposition for his ideas, based as they were on hypotheses tested by empirical evidence rather than ideas generated by a particular belief system or religion. The generally accepted theory at the time, in accordance with a close reading of the Bible, held that the world was once covered by an ocean, maybe caused by epochal events such as Noah's flood and that the general retreat of the waters of the ocean had revealed the continents. When the ocean receded, four distinct groups of rocks would become visible: primary rocks, mostly in the high regions and formed before life started to appear. Going through transitionary rocks and secondary rocks we would end up with alluvial rocks that were formed by floods and volcanoes, the more recent events. Fossils would only appear in the fluvial

---

[4] Hutton, J., 1788

rocks and in some in the transition rocks. The analogy of this explanatory system with the phlogiston theory come to one's mind, as it also could, even while being wrong by today's standard, explain a considerable number of important observable phenomena. In this case, the fact that the different rocks could also be conveniently located in the Bible's time frame helped no doubt to make this the accepted theory at the time. Hutton's view of an Earth much older than the biblical 6000 years brought him criticism from the Irishman Kirwan, who accused Hutton of making the Earth eternal, whereas in fact he had said that only with respect to human observation, the Earth had no end or beginning. In Hutton's view there was no beginning, no direction, and no end, and many of his critics could simply not accept such a directionless theory.[5] In his book version of the 'Theory of the Earth'[6] of 1795 Hutton addresses this and many other criticisms made at the time. Unfortunately, this, and the fact that his writing is at times rather convoluted and hard to follow, makes the book far less readable than the original lectures.

James Hutton also had a deep understanding of the link between the organic carbon cycle as exposed in the first two chapters of this book and geology. This is particularly visible in his discussion of coal. He claims that '*According to my theory, the strata of this earth are composed of the materials which came from a former earth; particularly these combustible strata that contain plants which must have grown upon the land.*' Coal was formed out of dead plant material. One of

---

[5] See the elegant 'Time's arrow, time's cycle: Myth and Metaphor in the discovery of geological time' by Gould, S., 1987. In the book Gould argues convincingly how Hutton's view of geological deep time and the recurrent cycle developed. In fact, he argues that Hutton may have developed this theory first, before he gathered some of what is now considered the crucial geological evidence, the discontinuities and granite intrusions.

[6] Hutton, J., 1795

the main tenets of his theory was that heat was a major factor in shaping the Earth. In the case of coal, he imagined that heat would have transformed the layers of organic material into coal. In this, his theory was still fully phlogiston based. *'Fire, and the consumption of phlogistic substances, is a great and necessary operation in the oeconomy of this world. There is constant fire in the mineral regions; fire which must consume the greatest quantity of fuel; the consolidation of the loose materials, stratified at the bottom of the sea, depends upon the heat of that fire; and the permanency of the land of this earth, above the surface of the sea, depends upon that consolidation of the strata, and upon the great masses of stone which had been in a melted state in the mineral regions.'* While the writing is rather cumbersome and characteristic of Hutton's style, it is, however, easy to read into this statement that fire not only produces the coals, but also functions as a source of energy for shaping the Earth. He claimed that this was one of the reasons why coals were absent under mountains and mostly found under flatland. In the first case they would have been burned up to provide the heat required to lift up the mountains. This, among other issues, brought, once more, Hutton considerable criticism from the Neptunist (catastrophist) Kirwan, who (correctly) held that there was no oxygen in the subterranean areas that would be able to keep the combustion of the coal going. However, Hutton held the rather complicated view that oxygen was not involved in producing the heat by combustion, but that it was the phlogiston that was released and produced the heat. It could be that Hutton was not a well-trained chemist, although being a good friend of Joseph Black he would have had a decent understanding, or that he did not really understand Lavoisier's theory. More likely, his thinking was still very much ingrained in the phlogiston theory and consequently he reasoned, using phlogiston as an energy carrier rather than gas, apparently in an ad hoc way to keep up his theory.[7] Whatever the exact reason, it would be many years before Wegener's form of plate tectonics (continental drift) would solve the issue of what energy drives the formation of continents and mountain ranges.

Hutton, through his theory of the Earth that combined uniformitarianism and heat, postulated the geological cycle as the key factor shaping the Earth. Uniformitarianism is the principle that the same processes that erode and shape the Earth occur and have occurred in the present and past. His views on carbon and coal suggest that he was also able to link the decomposition with the photosynthesis discovered by van Ingen-Housz in a first version of an organic carbon cycle (Chapter 5). Hutton's ideas were popularized by John Playfair, one of the other Edinburgh luminaries, who wrote 'Illustrations of the Huttonian theory of the Earth'.[8] It was through this book rather than the cumbersome writing of his own books that the public and other scientists finally became acquainted with Hutton's thinking.

Eight months after Hutton's death, Charles Lyell was born. He, more than any other, managed to convince the catastrophist, or biblical geologist to throw in the towel. Charles Lyell was trained to be a lawyer but had developed a keen interest in geology during his studies at Oxford. His 1830 book 'Principles of Geology, Being an Attempt to Explain the Former Changes of the Earth's Surface, by Reference to Causes Now in Operation'[9] did everything there was to be done to settle the issue whether the Earth was shaped continuously

---

[7] Kuhn, T., 1970 to see how scientists very often try to find ad hoc reasons to deal with anomalies to their theory or paradigm.

[8] Playfair, J., 1802. He also wrote a biography after James Hutton had died 26 March 1797, Playfair, J., 1805.

[9] Lyell, C., 1830-1833

operating dynamical forces or whether recovery of biblical floods shaped the continents. The book contained a wide review of studies that were cast and interpreted in the Huttonian theory. It would become the Bible of the geologists for the next 100 years. It shaped also the thoughts of Charles Darwin, who brought the book on his famous trip on the Beagle to the Galápagos Islands from 1832–1836, albeit with the warning not to accept all the views.[10] By the mid nineteenth century the idea of an Earth shaped by geological forces was firmly in place.

But what about the role of carbon and the carbon cycle? De Saussure had shown how plants take up water and CO_2 and turn this into sugars and oxygen (see Chapter 5). Importantly, he also discovered that oxygen was used to decompose plant material and that the ligneous (woody) component of the residue would thereby increase. This makes the cycle complete as was seen by Ébelmen who described this in 1845 as 'carbon rotation' and thought it was 'an admirable law of nature'.[11] Jacques Joseph Ébelmen was a professor at the 'École des mines' in Paris and Director of the Royal Works of porcelains in Sèvres, France (Figure 4.2), and was, it appeared fairly recently, the first to suggest that past changes in the carbon cycle could have an impact on the atmospheric concentration of 'carbonic acid' and thus on the climate of the Earth: 'many circumstances nonetheless tend to prove that in ancient geologic epochs the atmosphere was denser and richer in carbonic acid and perhaps oxygen, than at present. To a greater weight of the gaseous envelope should correspond a stronger condensation of solar heat and some atmospheric phenomena of a greater intensity.'[12] In the 1820s the French geologist

[10] Desmond, A. & Moore, J., 1991
[11] Galvez, M. & Gaillardet, J., 2012
[12] Ébelmen, J., 1845. Cited in Bard, E., 2004.

Figure 4.2 Jacques Joseph Ébelmen.

Adolphe Brongniard had laid some of the groundwork for this view. He realized that the large amount of coal deposits seen in the sedimentary record during the Carboniferous period would have been able to draw carbonic acid from the air. The naming of the Carboniferous is not coincidental; it means 'coal bearing' from the Latin words *carbo* and *fero*. It spans a period of 60 million years from the end of the Devonian Period 358.9 million years ago (Myr) to the beginning of the Permian period, 298.9 Myr.

He also firmly supported the view that coals were of biotic origin, something we have seen Hutton believed as well, but his critics, such as Kirwan, did not.[13] The importance of the work of Brongniard lies in the fact that burial of organic matter was by

---

[13] Desmond, A. & Moore, J., 1991

that time accepted as a means of controlling atmospheric CO_2 concentrations, and vice versa the levels of oxygen. We are getting close to the establishment of a geological component of the carbon cycle.

For the carbon cycle to function, the carbon locked up in the coal and ocean waters and sediments should not only be buried, but it should also be able to come back to the atmosphere. The Swiss physician and lecturer in chemistry Alexander Marcet had developed a device to sample sea water in 1819. Three years later he had identified calcium carbonate in a sea water sample: '*I found it to consist of … a portion of carbonate of lime [$CaCO_3$]. The presence of this last substance in sea-water, in a state of perfect solution, being, I believe, a new fact…*'[14] In another important development, a young French engineer, Jean Baptiste Boussingault, had started in 1832 to measure the composition of the gases and fluids that escaped from cracks in volcanoes. He established that 90% of the gas volume was composed of carbonic acid. With the smelly hydrosulfuric acid (the smell of rotten eggs), this made up the bulk of the gases coming from volcanoes. He suggested that in the formation of carbonic acid, heat and interaction with siliceous material would play a role (see Chapter 3; by heating lime, Black was able to obtain fixed air). Boussingault started to realize that not only there was an organic carbon cycle, that determined the composition of carbonic acid in the atmosphere, but that there was also an interaction with the deeper rock layers of the planet that played a role in determining the amount of CO_2 in the atmosphere.[15]

To be able to achieve a sort of balance or equilibrium, a form of taking out CO_2 of the atmosphere was needed. Understanding this

[14] Marcet, A., 1822
[15] Galvez, M. & Gaillardet, J., 2012

process was achieved by Bisschof, a student of Abraham Werner who was one of Hutton's loudest opponents. He realized that carbon dioxide played a key role in the dissolution (weathering) of feldspars in granite. This was largely due to the acidic character of water as a result of dissolution of the CO_2 in water. A parallel line of work identified the formation of kaolin through the dissolving effect of carbonic acid on granites. This effect was important for porcelain production. Ébelmen must have been aware of this, given his position at the Royal Porcelain Works in Sèvres. He suggested that silica and CO_2 would react together to form carbonates and that the minerals of magnesium and silicate would find their way into the ocean to be deposited. He did not yet have the benefits of the periodic table to describe everything in these precise chemical terms but he no doubt knew the elemental properties of silicas. He therefore established the geological importance of the process of weathering which is nicely expressed by Robert Berner, who recently credited him as the prime discoverer of the geological carbon cycle: 'The work summarized in his classic paper (Ébelmen, 1845) discusses in detail all of the major processes that affect the levels of atmospheric CO_2 and O_2, and his analysis, based on present-day knowledge, is complete and contains no omissions or reasoning errors.'[16]

Almost a hundred years later Harold Urey would write down the governing reactions as $CaSiO_3 + CO_2 \leftrightarrow CaCO_3 + SiO_2$. The first term, $CaSiO_3$, represents a simple igneous rock mineral called Wollastonite. These rocks usually also contain other elements, such as magnesium (Mg) or Iron (Fe), but that need not concern us. Urey (1893–1981) was famous for his discovery of the Deuterium isotope of Hydrogen for which he was awarded

---

[16] Berner, R., 2012

the 1934 Nobel Prize in Chemistry. He was also involved in the Manhattan project leading to the first nuclear bombs used on Hiroshima and Nagasaki and later in his career became interested in how atmospheres formed in planets. He would become particularly interested in how the composition of the atmosphere would affect life. One of his students, Miller, produced the famous Miller–Urey experiment that showed how from the prime components of the presumed atmosphere 4–3 billion years ago and electric discharges, simple amino acids could be produced. We are here most of all concerned with the role of CO_2 and it is interesting to hear what he has to say about that, from his 1952 paper: 'As carbon dioxide was formed it reacted with silicates to form limestone. Of course, the silicates may have been a variety of minerals but the pressure of CO_2 [at the Earth] was always kept at a low level by this reaction or similar reactions just as it is now. Plutonic activities reverse the reaction from time to time, but on the average the reaction probably proceeds to the right as carbon compounds come from the earth's interior, and in fact no evidence for the deposition of calcium silicate in sediments seems to exist.'[17] The Urey equation is a cornerstone of our understanding of the geological carbon cycle, and sometimes in the climatological context referred to as the geological thermostat.

It is worth diving into this a little bit further. Si is the chemical shorthand for silicon, silicates are the molecules that form when it is oxidized. The Urey equation is a simplification in the sense that the weathering of the silicas works only in the presence of water that dissolves the carbon dioxide. Plants, which respire CO_2 and physically split rocks through root growth, may further amplify the weathering but water is essential. This is the beauty then

---

[17] Urey, H., 1952

of the geological thermostat: higher temperatures mean more rain (in principle the atmosphere can contain more moisture at higher temperatures), and thus the weathering can increase. This increased weathering in turn lowers the atmospheric concentration of CO_2 so that it can become cooler again. When it becomes cooler, the weathering decreases through less availability of water through precipitation and the CO_2 concentrations in the atmosphere will increase again, until the cycle repeats itself. This cycle it thought to operate at geological timescales of millions of years (so unfortunately it does not help to solve today's climate crisis). Weathering is one aspect of the geological carbon cycle. The provision of CO_2 through volcanic eruptions is the other. As shown by Boussingault this provides the source term to the equation that generates variability at geological timescales. The stability of our climate over the last million years or so is usually taken as evidence that this simple feedback of the geological thermostat has worked well to keep the Earth's climate in check. If the CO_2 source (the volcanoes) and the sink (the weathering) were out of balance by 10%, they could easily double the CO_2 concentration in the atmosphere in about 50,000 years, but no such large variations have been found. Other, long-term evidence suggests that the presence of sedimentary rocks of carbonates (the $CaCO_3$ part of the Urey equation) from the early stages of the Earth indicates that temperatures have generally stayed at such levels to allow water to occur in its main three forms: vapour, frozen, and liquid.[18]

The concept of the carbon cycle as we understand it today, as one of the Earth's key biogeochemical cycles appears to have been originated by the Russian Vladimir Vernadsky (1863–1945).

[18] Archer, D., 2010

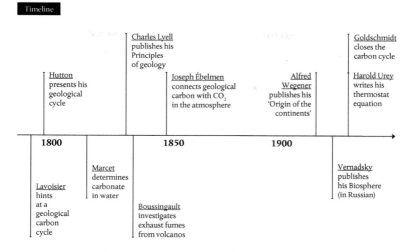

Figure 4.3 Timeline of the main events described in this chapter.

The word biogeochemical, though it may appear at first glance a rather clumsy construct, does indicate exactly what it is about: the interaction of geological with biological processes in the form of a cycle with continuously changing roles of carbon in specific parts of the cycle. In the next chapter we will focus on the role of the biological part, respiration and photosynthesis, but we also saw previously that Black produced his first fixed air by heating lime. These chemical processes involve the weathering and dissolution of minerals that we discussed above. It is now time to close the cycle by looking more closely at the geological part (see also Figure 4.3).

Rocks are formed by so called igneous processes. They form when hot, molten material inside the Earth crystalizes and becomes solid. Geologists identify two kinds: intrusive and extrusive rocks. The first form when the molten material, magma, slowly rises towards the surface and cools over many thousands or millions of years. The other form, the extrusive rocks, are the ones that we are used to see erupting from volcanoes or large fissures in the Earth's crust. These rocks cool almost instantly and have a different texture than the intrusive rocks. Hutton was already in 1887, in his famous excursion to Glen Tilt, looking for intrusive rocks that would allow him to substantiate his claim that heat was the main driving force on Earth. He indeed located several granite veins that intruded into the (older) schist formations. Granite was formed by his subterraneous heat (now known as long-sustained crystallization) that had forced its way into the sedimentary rocks. It could well be the force that raised the Earth!

Sedimentary rocks form when small particles sink through the water column down in the ocean and/or open water to the bottom (a process called sedimentation) or directly through the air on the land surface. During continuous sedimentation the weight of the overlying layers compacts the lower ones and ultimately forms sedimentary rocks. Sedimentary rocks form the bread and butter of geology as their location, composition, and structure tells us something about the environment at the time they were formed. Broadly there are three types of sedimentary rocks. The first category, clastic sedimentary rocks, form from the remains of physical weathering; they appear in the form of rocks such as breccia and sandstone. The second category, chemical sedimentary rocks, form when dissolved material from weathering precipitates in water; examples are iron ore, but also dolomites and some limestones. The third category consists of organic sedimentary rocks such as coals, dolomites, and limestones, and form from accumulated organic material from plants or animals. The third category is where the geological cycle interacts with biology.

Figure 4.4 shows the workings of the geochemical part of the carbon cycle. Continental rocks are being weathered by the action

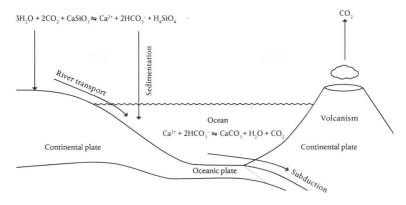

$$3H_2O + 2CO_2 + CaSiO_3 \leftrightarrows Ca^{2+} + 2HCO_3^- + H_4SiO_4$$

River transport

Sedimentation

CO_2

Ocean

$$Ca^{2+} + 2HCO_3^- \leftrightarrows CaCO_3 + H_2O + CO_2$$

Volcanism

Continental plate

Continental plate

Oceanic plate

Subduction

Figure 4.4 The geological carbon cycle containing the geological thermostat. The rate at which the sediments are buried and subsequently returned to the atmosphere through subducting plates that emit volcanic carbon dioxide then determines the net long-term balance of carbon dioxide in the atmosphere.
Redrawn after Dolman, H., 2019

of CO_2 and water. This is the Urey reaction written in a slightly more complex form, indicating the key role of water in the geological thermostat. The products of this weathering are the calcium and bicarbonate ions and silicic acid that are transported by rivers into the ocean. In this process other solutes such as magnesium and iron are also transported into the ocean, where the conditions may be such that they precipitate, thus forming a chemical sediment. In the ocean, the Ca is taken up by small algae and other unicellular lifeforms and form calcareous skeletons that eventually sink to the ocean floor and form the chalk stone we see for instance at the English coast near Dover. These are prime examples of the interaction of geology with biology. If it is not taken up by the unicellular organisms it will form calcite (calcium carbonate), the ground material for limestone.

Hutton had gone to great lengths to identify heat as the main force for lifting rocks. He was close, but could not have understood how the current thinking of plate movement would be able to explain such phenomena while also providing a mechanism for the carbon cycle to operate on geological timescales (often in excess of several hundred millions of years, timescales almost unimaginable to the human mind). This theory was proposed by Alfred Wegener, by origin a meteorologist. He had not been the first to realize that the continents of Africa, Europe, and the Americas seemed to fit together remarkably well, but he was the first to develop this into a full-blown theory of how continents moved. At the beginning of the twentieth century most geologists held the view expressed by Eduard Suess (the father of Hans Suess, see Chapter 9) that during the contraction of the once-warm Earth, the outer crust shrank a bit like the skin of an old apple and that uneven shrinking produced the continents and mountain ranges. But there were serious issues with that view, one of them being that you would expect mountains to be more evenly scattered around the Earth, rather than appearing as fixed, elongated mountain ranges such as we see in the Himalayas, Alps, and Andes. Another was that it was now known, since radioactivity was discovered in 1896 by Henri Becquerel, that radioactive decay in the inner Earth would continually supply heat and this would tend to work against the assumed cooling. In 1912 Wegener published his first paper on his theory that the continents were formed by drifting apart, and mountain ranges by colliding tectonic plates. In 1915 he published the first version of his book the 'Origins of continents and oceans'[19] that he updated in 1929. In it he

---

[19] Wegener, A. & Biram, J., 1966

suggested that the continents had once formed a single big conti-
nent, Pangaea, that was slowly drifting apart into the continents
we now know. Continents would not only move vertically as on
Suess' shrinking crust, but also horizontally. He also suggested
that the continents were of lighter (less dense) material than the
mantle and thus were able to drift. The uptake of his theory was
slow, as most geologists would not buy such a theory that in their
view lacked the precise mechanism or force that would make the
continents flow. At some stage, it was even called 'Germanic pseu-
doscience'.[20] It would take another 40 years and vicious battles
to get the theory accepted, 40 years in which paleomagnetic data
finally provided evidence of sea floor spreading in the oceans. It
would also take the same 40 years for a meaningful theory to be
provided that included the driving force for the subduction of
continental plates beneath oceanic ones. The subducting plates,
like the one in Figure 4.4 where a slab of sedimentary material
dives under the ocean crust, generate earthquakes and volcanoes such as near Indonesia, part of the well-known Ring of Fire
around the Pacific. The subduction of sedimentary plates, with
the corresponding increase in pressure and temperature, eventually produces the gases for the volcanoes. And thus, geology and
the carbon cycle become intricately linked though Wegener's
theory of continental drift, or as it is presently named, plate
tectonics.

Back to Vernadsky, who achieved his wider fame 30–40 years
ago before the now popular Gaia theory, developed by Love-
lock and Margulis appeared. This theory describes how the Earth
behaves as an organism, keeping its temperature through the

[20] Powell, J., 2015

operation of multiple biogeochemical feedbacks in which life is both a result and cause of the climate. They were at the time, ignorant of the work of the Russian scientist. Vladimir Vernadsky is one of the founders of the concept of the biosphere, a term originally coined by Eduard Suess. In the mind of Vernadsky[21] this went a considerable amount further than the total sum of the sedimentary rocks formed by organisms as suggested by Suess. Instead, he included composition, dynamics, and energetics in his definitions, which further held that the biosphere was a layer on the Earth several kilometres deep, including bacteria, the hydrosphere, and the lower layers of the atmosphere. Vernadsky thought that the continuous movement of materials in the Earth's crust was facilitated or mediated by organisms. Solar and chemical energy provide the primary source of energy for living systems that drew non-living matter (minerals) into continuous circulation. These are the key concepts of a biogeochemical cycle where organisms powered by the Sun perform large-scale migration of material in the Earth's crust, or in his view, the biosphere. He reasoned that life was inextricably tied to the biosphere, and that this was no accident: it allowed the biosphere to function. The origin of photosynthesis was considered by him as one of the critical evolutionary steps needed to change material from autotrophic to heterotrophic organisms (Chapter 5). In short, the biosphere is a fully connected system of live and non-living matter that has organized itself and thereby conditions to a large extent its own environment. Life is not only driven by geology, but importantly also acts as a geological force itself, shaping geology (see also Figure 4.5).

[21] Vernadsky, V., 1926

Figure 4.5 Image of a
seated Vladimir
Vernadsky.

While it took the Western world some time to accommodate
Vernadsky's work—the fact that several of his key papers were in
Russian would not have helped the uptake—around 1920 a scientist from Norway, Victor Goldschmidt, also began to be interested
in geochemistry. In a thorough analysis of 40 letters between him
and Vladimir Vernadsky between 1913 and 1939 the importance of
Vernadsky's thinking on the development of Goldschmidt's ideas
becomes clear.[22] Goldschmidt, now considered the father of modern geochemistry, with an important annual conference named
after him, started his career in the mineralogy and petrology of
metamorphic rocks in the Kristiana area in Norway. He became a

---

[22] Müller, A., 2014

professor at the early age of 26 at the University of Oslo and director of the Raw Material Laboratory, raw materials having become
increasingly important after the First World War. The analytical
possibilities that came with the lab led him on the path to geochemistry with as result a groundbreaking series of nine volumes
entitled '*Geochemische Verteilungsgesetze der Elemente*', or 'The geochemical laws of the distribution of elements'. His main aim was
to establish the distribution of the elements in the Earth's crust
and in meteorites and what would be the cause of that distribution: '*It is conceivable that the original state of the Earth was a homogeneous
or nearly homogeneous mixture of the chemical elements and their compounds. Today, however, the Earth is in a far from the homogeneous state.
The material distribution within the Earth has by no means reached a final
state of equilibrium; we observe instead an active redistribution of matter
and energy. The processes which have resulted in the inhomogeneity of our
planet and still contribute to the migration of material I would summarize in
the expression "Der Stoffwechsel der Erde" (The metabolism of the Earth).*'[23]
He developed Goldschmidt's rule of diadochic substitution, which
states that elements with ion ratios within 15% of each other could
replace each other in a crystal lattice. During his period in Oslo,
Goldschmidt was visited by Vernadsky in 1927, an indication that
they were in close contact and that he was indeed familiar with his
work.

Two years later, Goldschmidt moved to Göttingen, Germany.
He was visited again by Vernadsky to discuss various geochemical issues—amongst others, the carbon dioxide content of the
oceans. Theoretically and conceptually, Vernadsky was always
some steps ahead of Goldschmidt, but Goldschmidt's analytical

---

[23] Goldschmidt, cited in Müller, A., 2014.

techniques provided him with a range of chemical data that put him well upfront in empirical analysis. Importantly, his methods allowed him to draw quantitative conclusions. He calculated, for instance, the amount of material weathered by geological processes and the amount of sediments produced. He applied this method also to the carbon cycle, and this work may be considered one of the first quantitative expressions of the carbon cycle. Goldschmidt's attempt qualifies as the first complete quantification of the carbon cycle. He does acknowledge his debt to Vernadsky in his 'Drei Vorträge über Geochemie', published in 1934:[24] 'Es its vor allem das grosse Verdienst von W. Vernadsky den Umfang und die Bedeutung biologischer Vorgänge in de Geochemie mit allem Nachdruck betont zu haben.'

Goldschmidt links the organic carbon cycle with photosynthesis and respiration and the geological cycle of coal and limestone with the concentration of CO_2 in the atmosphere. His result is shown in Figure 4.6. The Swede Högbohm, collaborator of Arrhenius, some thirty years earlier, had estimated that about 650 Pg C was residing in the atmosphere. This is close to current estimates of the preindustrial carbon content of the atmosphere, at 589 Pg C. He also figured out that weathering flux would be almost equal to the fossil fuel flux which he estimated at 0.14 Pg yr^{-1}. He suggested that variability in atmospheric concentrations would be most likely be caused by volcanoes exhuming CO_2. The US geologist, Chamberlain, also thought around the beginning of the twentieth century that CO_2 variations could be large enough to plunge the Earth into an ice age, but he maintained that it was not volcanic activity that was the main cause,

[24] Goldschmidt, V., 1934

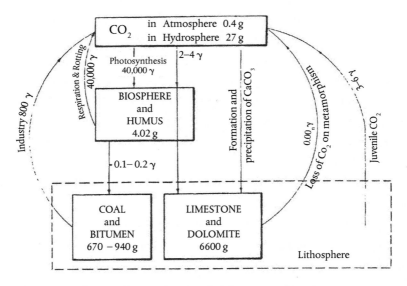

Figure 4.6 The geochemical cycle of carbon according to Victor Goldschmidt (1 Υ = 10^{-6} gram per year).
From Mason, B., 1992

but the weathering of rocks and the exposure of these to the atmosphere.[25]

Goldschmidt takes an important step in putting up this picture of the carbon cycle that not only shows the reservoirs but also the fluxes that link them. His estimates of the sizes of the reservoirs are eerily close to today's estimates; atmosphere (556/590 Gton C); ocean (37554/38700 Gton C); biosphere (306/500 Gton C); soils (1390/1500 Gton C); limestones (9,180,000/65,000,000 Gton C). Importantly he also linked the organic balance with photosynthesis and respiration with the geological balance of weathering. Around the same time, the Dutchman Baas Bekkering in his book

[25] Fleming, J., 1998

'Geobiology' presented a similar cycle picture, albeit structured with the redox potential of the components, i.e., the sugars as most reduced compounds at the top and CO_2 as the most oxidized at the bottom. Clearly, thinking of biogeochemical cycles as a series of linked reactions was becoming prevalent in the years just between the two world wars.

Today's perception of the geological carbon cycle as shown in Figure 4.7 draws heavily on the work of Hutton, Ébelmen, Vernadsky, and Goldschmidt. The Urey reaction is central to the long-term geological cycle. Remember we are talking about processes that take many, many millions of years. It describes how calcium silicate (the mineral wollastonite) takes up CO_2 (in a watery solution) and produces silicates and calcium carbonates that can form limestones when organisms that take up the calcium carbonate sink down to the ocean floor. The return

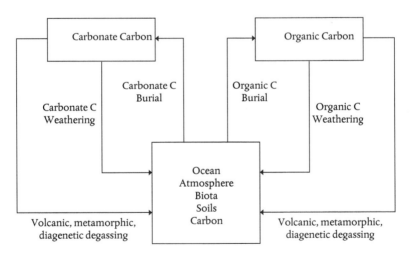

Figure 4.7 Geological carbon cycle.
Redrawn after Berner, R., 1999

flow of carbonates then goes through weathering and outgassing via volcanoes. The right-hand side of the picture denotes Hutton's coals. Similarly, these form from biospheric products in the big box containing the biospheric carbon in the ocean and land. This box links of course to the respiration and photosynthesis processes we will deal with in Chapter 5. In the words of James Hutton: *'no vestige of a beginning, - no prospect of an end'*.

CHAPTER 5

CARBON DIOXIDE AND THE VEGETATION OF THE EARTH

We trace the discovery of photosynthesis, the process in plants that uses carbon dioxide and water to produce sugars and oxygen. If this process did not exist the world would look rather different, and animals would most likely not exist. Jean Baptista van Helmont and Joseph Priestley, well known for their work on carbon dioxide, as described in the previous chapters, discovered that for plant growth water was needed. In the early eighteenth century Stephen Hales had discovered that plants transpired water and that they were thus capable of gaseous exchange with the atmosphere. Joseph Priestley put pieces of plants in water and noticed that bubbles of air started to form around the leaves, but he failed at the time to link this to the production of gas from the leaves through photosynthetic production of oxygen. He did, however, notice that candles in a flask burned longer when small plants were added to the flask. In a frantic summer of experimentation in England, the Dutchman Jan van Ingen-Housz discovered that light was an additional essential component. The final pieces of the puzzle were laid by Jean Senebier who realized that leaves were in fact small laboratories where Joseph Black's 'fixed air' (carbon dioxide) was transformed into 'pure air' ('dephlogisticated air', later known as oxygen). He was one of the first, if not the first, to realize that, next to the slow geological carbon cycle, a fast carbon cycle existed, where biomass is produced that is ultimately respired to produce water and CO_2, and then..., the process of producing biomass through photosynthesis begins again.

At the beginning of the nineteenth century the basic elements of the mystery of photosynthesis were unravelled. Théodore de Saussure managed to present an even more complete picture of photosynthesis by further careful experimentation. He was in fact one of the first scientists to perform what we now call an artificial fertilization experiment, where he exposed plants to air enriched in carbon dioxide. He determined they grew faster... But what happened exactly in the green parts of the plants was still unknown. Somehow it appeared that water, CO_2, and light miraculously produced oxygen and carbohydrates (sugars). It took a while to unravel this and discover what exactly happened. The first answer came from photosynthesizing microbes, more precisely the purple bacteria of Kees van Niel who understood what happens when light hits them. Van Niel discovered that the microbes produced oxygen from water like they did sulfur from the smelly hydrogen sulfide. Plants and microbes essentially work according to the same mechanism! In the case of green plants this means that the oxygen unequivocally comes from the water and thus not, for instance, from the CO_2. Light was somehow harvested and used to split the water. Robin Hill isolated the main parts of the cell that played the key role in photosynthesis, the chloroplasts. These small organelles contain chlorophyll, the absorbing green pigment in plant cells. Hill also established that in the dark, photosynthesis stopped. The first step of photosynthesis was understood.

But what about the second step, how is carbon dioxide then fixed into sugars? It was up to Calvin and his co-worker Benson to cleverly use isotopes *to determine how CO_2 was incorporated into sugars. This process is now known as the Calvin–Benson cycle. The enzyme that plays a critical role here is Rubisco, probably the most important enzyme on the planet, at least from a biogeochemical perspective. Evidence for its effectiveness lies in the fact that the Rubisco-based system has changed very little from the first time it appeared on the planet some 3 billion years ago.*

Scientific discoveries rarely stand on their own. There often is a murky background of previous studies, unexplained phenomena, strange results, and, quite regularly, similar, and even simultaneous, experiments performed by several people. In all this complexity we tend to identify that person who made that single important and precisely defined conceptual jump. That historical process is almost by definition fraud: fraud with the errors of neglect of other experiments and people who also contributed, albeit only small steps. These forgotten experiments and unsung heroes might appear less crucial in the whole story, but it is important to remember that the perspective of a single discovery depends very much on the fact that we look back at the history of science with the knowledge of today. With that knowledge we tend to emphasize history that fits into a somewhat logical, often linear picture of development. Arguably, developments that contributed more to the outcome we now know to exist are considered as more important than the multitude of blind alleys that most likely were also part of the discovery process, if not the majority.

The initial cast of characters identifying the role of plant photosynthesis on planet Earth is, however, remarkable, like that in

Chapter 3. Along again come Jean Baptista van Helmont, whose willow experiment first proved the role of water in plant growth (although he 'forgot' the role of leaves) and Joseph Priestley, who discovered that when a small mint plant was put into a flask a lighted candle would burn longer. To call them discoverers of photosynthesis would go a little too far. They did however contribute much to setting the scene. Oxygenic photosynthesis, as is the official name, is the process whereby plants under the influence of light produce sugars and oxygen from just two ingredients, water and carbon dioxide. Van Helmont correctly identified water as a critical component and Priestley hinted at the existence of oxygen that replaced noxious air (containing carbon dioxide) when a small plant was put inside a flask with a burning candle. However, the honour to be called the discoverer of photosynthesis falls to the Dutchman Jan van Ingen-Housz (1730–1799). He identified light as the key requirement and realized that the process took place in the forgotten leaves of van Helmont's willow tree.

As we have seen in the previous chapter, the history of the discovery of carbon dioxide is very much tied to that of oxygen. Priestley and Lavoisier both did experiments that identified not only the generation of oxygen from mercury oxide, but also realized that air was being made 'noxious' when animals breathed in it. Noxious air was in fact air that contained carbon dioxide, and little oxygen, and mostly nitrogen. Normal air contains mostly nitrogen (78%), oxygen (20.9%), and carbon dioxide (now slightly above 0.04%); other trace gases make up the rest. Plants or mice that were put into sealed containers mostly died quickly. Joseph Black was interested in purifying this noxious air that was given off by life.

Combustion, the use of oxygen to burn a substance, either through a chemical process or through respiration of life, and the reverse, the production of oxygen through photosynthesis, are tightly coupled in the Earth system through what is called the fast carbon cycle. The long-term stability of the levels of oxygen (at geological timescales) is another matter, which we just encountered in the previous chapter, but let us for now concentrate on that faster cycle. In that process we can safely assume that oxygen levels are constant at today's levels of about 20.9%.

Figure 5.1 shows the key relations between, on the one hand, the production and consumption of oxygen and, on the other hand, the production and consumption of carbohydrates, the official name for what we would call sugars. Carbohydrates are defined as products all containing carbon, hydrogen, and oxygen, with

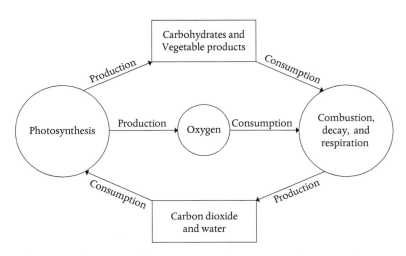

Figure 5.1 Schematic of the interaction between production and consumption of oxygen and the production and consumption of energy.
Redrawn from Nash, L., 1957

sugar being one of them. We have already seen in the first chapter that combustion and respiration by life forms (animals and plants) produce carbon dioxide. This is the lower right-hand side of the figure. In the upper right-hand part energy is produced by these combustion and respiration processes. The left-hand side of the figure represents one of Nature's big inventions: photosynthesis. Under the influence of light, water and carbon dioxide are converted into carbohydrates, making the cycling almost self-sustaining. The waste products of energy consumption are being reused to reproduce the fuel in the form of carbohydrates that, when burned, produce energy and new waste products, and so the cycle goes on and on.... In fact, while this cycle is normally called the fast carbon cycle, one could also call it life's fundamental cycle, as all forms of life depend on it. The producers of carbohydrates are called autotrophs, as in principle they can be self-sustaining by producing their own fuel like plants. Other life forms such as animals or microbes that lack the ability to photosynthesize are called heterotrophs. The latter depend on photosynthesizing life forms to generate the energy they need to maintain and reproduce.

Let us go back to van Helmont's willow. He established that water could be transported into a tree by transmutation, this being the way he looked at it with his alchemic background that still guided him. A few years later Robert Boyle executed several experiments that basically confirmed that plants could grow on water. He performed, for instance, one experiment in which he grew plants on water alone.[1] It is probably fair to say that around that time, the end of the seventeenth century, the experimental

---

[1] Nash, L., 1957.

philosophers such as Boyle did not think that the atmosphere played a role in plant growth. They very much thought that water was being transmuted into plant material. That picture began slowly to change when the Englishman Nehemia Grew and the Italian Marcello Malpighi started to use the early microscopes to study plants, trees, and in particular the surface of leaves. These first microscopes were developed by Robert Boyle's scientific partner Robert Hooke and the Dutchman Anthony van Leeuwenhoek. Malphighi would become most famous for his 1661 discovery of pulmonary capillaries and alveoli, the tiny air sacs in the lungs responsible for exchanging gases such as oxygen and carbon dioxide with the blood. His early drawings of the structure of the lungs are true miracles of precision and interpretation. Nehemia Grew, also using the newly developed microscope, saw that 'the skins of at least many plants are formed with several orifices or passports, either for the better avolation of superfluous sap, or the admission of air'.[2] The use of microscopes to study previously unseen aspects of the anatomy of plants is one of the examples of two scientists working independently and arriving at similar conclusions.[3] Grew was proposed as member of the Royal Society by Robert Hooke in 1671. Grew's essay had only just been printed when Malpighi's draft was submitted. On a day, just before Christmas (21 December) in 1671 Malphigi's work was also read in the Royal Society. That day stands in history, not only because of this momentous occasion, but also because on the very same day, Isaac Newton was proposed for membership of the Royal Society.[4]

Neither Grew not Malphighi related the structures they saw to the flow and transport of water, or sugars for that matter, in

[2] Grew, H., 1682
[3] Hunter, M., 1982
[4] Arber, A., 1942

plants. That insight was developed by the Englishman Stephen Hales about half a century later. In his 'Vegetable Staticks', published in 1727, he did acknowledge his debt to his two predecessors: 'But our countrymen Grew and Malpighi were the first, who, tho'in very distant countries, did nearly at the same time, unknown to each other, engage in a very diligent and thorough inquiry into the structure of the vessels of plants; a province, which till then had lain uncultivated.'[5] He goes on to say that if these two scientists would have applied the methods he used, i.e., the new hydrostatic experiments, they would have probably got similar insights as he obtained. He used the term 'statick', or static research, to mean weighing or gravimetric analysis and the study of the forces of fluids by manometry (measuring the pressure). His work on plants cleverly made use of manometers that were tied to trunks and shoots (Figure 5.2). Hales also performed weighing experiments, somewhat reminiscent of van Helmont's experiments, where he covered the soil in which a plant stood in a closed environment to determine how much moisture they lost or gained. In that way he determined the main patterns of the flow of water in plants and, importantly, that they exchanged that water with the atmosphere. He was one of the first to use control experiments, which was a major innovation in scientific methodology. His work on blood pressure, as published in the companion book 'Haemostaticks' from 1733 but containing work that he had done much earlier on a variety of dogs and horses, made him justifiably known in medical circles as one of the great scientists of his time.

But for our story, more important is that, while he was one of the first plant physiologists, Hales also experimented with the impact

---

[5] Hales, S., 1727

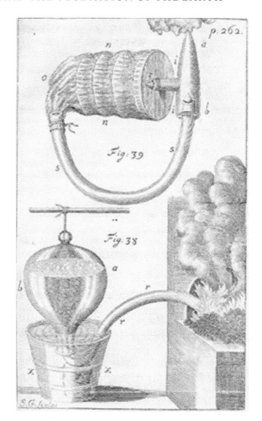

Figure 5.2 Drawing out of *'Vegetable Staticks'*, the book in which Stephen Hales recounted his discoveries about the flow of water in plants and hints at the interaction of air with plants. Here is shown his innovative device, the pneumatic trough, which allowed him to study the air that would come from fermentation or decomposition.
From Chapter 7 of Hales, S., 1727

of plants on air. It is in this context that he performed his experiments with the pneumatic trough (Figure 5.2). The pneumatic trough was a critical instrument in the development of science, as it allowed for the first time for the collection and handling of gases.[6] Hales remains a little vague about the air and its properties he thus collected in his many experiments, even though his Chapter 6, *'The analysis of air'* is by far the longest of the book. He

[6] Nash, L., 1957.

quite clearly struggles to get away from the alchemic jargon but does establish a link between vegetation and the atmosphere. He does realize the role of leaves: *'we may therefore reasonably conclude that one great use of leaves is what has long been suspected by many, viz. to perform in measure the same office for the support of vegetable life, that the lungs of animals do; plants very probably drawing thro' their leaves some part of their nourishment from the air....'.* What makes him, maybe not the discoverer, but certainly one of the groundbreaking predecessors of the idea of photosynthesis is the next statement: *'And may not be light, by freely entering the expanded surfaces of leaves and flowers, contribute much to the ennobling principle of vegetables.'* He refers to Isaac Newton who had suggested that gross bodies and light could be convertible into each other. His critical experiments, again using controls, involved putting plants into a closed-off flask of air and measuring the reduction of the volume of the air by 1/7th as a result of plants respiring (*'reduced to a fixt state'*, i.e., converted into fixed air, carbon dioxide). Leaving the system for a few months over the winter, in the spring he put a new plant into the flask that died quickly. When he put a plant into the control flask, in which no plant had previously been put, he observed that it continued to live. He thus discovered that the air degradation in the first flask was wholly due to the interaction of the plant with the flask's atmosphere. What precisely this interaction consisted of, he did not know, but he had made the link of gaseous exchange of plants with the atmosphere and he had hinted at the importance of light.

In Chapter 3 we encountered the towering scientific figure of the times, Joseph Priestley. He put candles in flasks and noted that the candles burned longer when a sprig of mint was also put into them. He struggled to reach a firm conclusion of these experiments, i.e., whether the air was being made bad because of the plant's respiration by producing carbon dioxide or improved by making oxygen through photosynthesis. It appeared that in his experiments he mistakenly oversaw the workings of small algae, green matter, producing oxygen through photosynthesis, that formed in his flasks.[7] He had, however, noticed that when he put pieces of plants in water, bubbles of air started to form around the leaves, but he failed to link this to the production of gas from the leaves through photosynthetic production of oxygen. He had observed also that light appeared to play an important role.

Jan van Ingen-Housz did see the relevance of Priestley's 'green matter'. Through a series of experiments, he was able to show that the green matter was an active plant able to photosynthesize. Van Ingen-Housz published his main work as 'Experiments Upon Vegetables, Discovering Their great Power of Purifying the Common Air in the Sun-shine, and of Injuring it in the Shade and at Night' in 1779.[8] That he published in English is not a real surprise. Van Ingen-Housz was a man of the world. With an original degree in Medicine from Louvain, he had worked at the universities of Paris, Edinburgh, and Leiden on chemistry and physics.[9] He was known as an expert in the inoculation of smallpox. After having inoculated most of the Austrian royal family, i.e., the Austrian archdukes Maximillian, the younger brother of the emperor Franz-Joseph, and Ferdinand, and the archduchess Theresa, he was rewarded with a lifelong income and various additional gifts that made him, by and large, financially independent. In 1771, after having travelled around

---

[7] Nash, L., 1957.
[8] Ingen-Housz, J., 1779
[9] Gest, H., 1997

Europe and inoculated various other members of the Austrian imperial family, he went to London where he was admitted to the Royal Society.

For our story his period in England in the summer of 1779 is important. While on leave from Vienna he performed an incredible amount of more than 500 experiments in just three months. Still working with the phlogiston theory, he concluded that plants emit dephogisticated air (oxygen) under the influence of light. When light was absent, the plant in contrast made the air bad (i.e., with carbon dioxide). He was also able to show that it was primarily the leaves that performed this activity. Van Ingen-Housz considered this to be a transmutation in the leaves under the influence of light.

It now took men (yes, unfortunately mostly men) from continental Europe to further make progress in understanding. Jean Senebier, the son of a merchant and pastor, was a Swiss from Geneva and a keen experimenter. It is remarkable, by the way, how many of the scientists at that time were also pastors. In 1754, Charles Bonnet, one of Senebier's earlier colleagues in Geneva, had discovered that leaves would produce no bubbles in water when they were put in distilled water (i.e., water free of fixed air, carbon dioxide). Van Ingen-Housz had made the same observation, but he also discovered they would produce bubbles when they were placed in pump water that was considered richer in fixed air. He did not, however, conclude from that initially that leaves needed carbon dioxide, and it was Senebier who made that important conceptual leap. He performed several experiments where he investigated how much fixed air was needed to produce the dephlogisticated air. From these experiments he was able to conclude *The leaves are nothing but laboratories in which is prepared so much the more pure air as there is in the surrounding medium more of the fixed air that they elaborate.*[10] He had identified the link between the uptake of fixed air (carbon dioxide) and the production of dephlo- gisticated air (oxygen). It is worth emphasizing that at this stage the phlogiston theory from Priestley was still being used success- fully to explain the key properties of photosynthesis. Senebier went further showing that shredded leaves were also able to pro- duce dephlogisticated air, and that neither a whole plant (per Joseph Priestley) or a whole leaf as van Ingen-Housz thought, was needed to achieve the production of bubbles. He went further in showing that those parts of the leaves, skin, i.e., the non-green bits of plants, were not able to produce the dephlogisticated air.

In 1782 Senebier interpreted his results fully within the phlogis- ton theory. When plants were illuminated in air that contained phlogiston, they were found to emit dephlogisticated air (oxygen). As the gases then mixed in the dephlogisticated air, the remain- ing air takes up some of the readily available phlogiston in the air, and thus forms fixed air. Since this was known to be heavier than air it would fall towards the bottom. This was something already well known: in the famous Grotta di Cane, near Pozzuli, near Naples in Italy, first described by Pliny the Elder, geologically formed carbon dioxide produces a layer about 1 metre high at the bottom where dogs would quickly suffocate and go unconscious. Humans breathing at a height above the 1 metre layer experience nothing. Once the fixed air precipitated according to Senebier, it would get into contact with groundwater where it would dis- solve, to be taken up by roots and stems and transported to the leaves, as Hales had shown to be the case. In the leaves the light would act to release again the phlogiston from the fixed air and

---

[10] Nash, L., 1957.

the dephlogisticated air would be realized. The importance of this scheme is twofold here; first it provides a mechanism to purify air (remember Priestley's candle?) and secondly it provides a cyclic mechanism that, even though based on the phlogiston theory, looks remarkably like the fast carbon cycle, that involves photosynthesis and decomposition. Van Ingen-Housz and Senebier spent a good deal of time arguing against each other on details of this scheme, in particular whether it was the fixed air that was converted into pure air, and with van Ingen-Housz accusing Senebier of plagiarism, since he had repeated many of van Ingen-Housz's experiments (Figure 5.3).

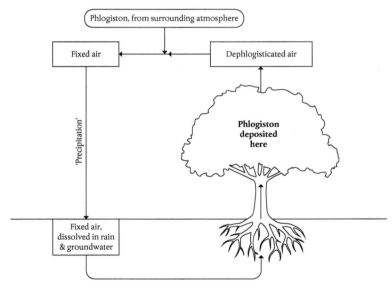

Figure 5.3 The conceptual scheme of Senebier; what we see is the fast carbon cycle in phlogiston perspective. See text for further explanation. Based on the original drawing from Nash, L., 1957

By 1796, after Lavoisier had redefined chemistry and overthrown the phlogiston theory of Priestley (see Chapter 3), van Ingen-Housz was now finally able, or mentally ready, to interpret his results and those of Senebier more clearly. He states '*as water itself is a composition of two airs, vital and inflammable, or oxygen and hydrogen, in which two substances, Mr Lavoisier found means to analyse water, and which analysis, as far as it regards the oxygen, I affirmed in my first volume on vegetables, to be performed by Vegetables, with the assistance of the sun even before Mr Lavoisier, as I think, published his Analysis*' and '*Besides this power of shifting carbonic acid from the air by attracting its oxygen and furnishing it with carbon, plants possess a most wonderful faculty of changing water itself into vital air, or oxygen; which I have maintained as early as 1779 (See my work on Vegetables).*'[11] Reading that particular paper one is surprised by its modern character, and the ease with which he has absorbed the new chemistry developed by Lavoisier. He also realizes that animals depend for the food on vegetables (they are heterotrophic as we have seen) and plants are dependent on the atmosphere and produce their own (they are autotrophic). Here is a man who has understood the first principles of our Figure 5.1! By the beginning of the nineteenth century the basic mystery of photosynthesis was thus resolved, and it was clear that both plants and animals needed oxygen to respire, the fuel being the carbohydrates produced by photosynthesis in plants.

The basic mystery was solved…, at least qualitatively. It required the quantitative approach of Nicolas Théodore de Saussure to finish the work and present a complete picture of photosynthesis. De Saussure was thoroughly versed in Lavoisier's chemical system and a meticulous experimenter. His experiments were presented

---

[11] Ingen-Housz, J., 1797

in his 'Recherches Chimiques sur la Vegetation' published in 1804. This is only about eight years after we saw the conversion of van Ingen-Housz to Lavoisier's system. De Saussure was able to show that water was important for plant growth. While van Helmont had thought that water was being transmuted in his willow experiment, de Saussure made this more precise. He also appreciated that even with the tiny amounts of carbon dioxide available in the air (we do count them after all in parts per million) plants were able to grow by using this. He performed what we now would recognize as an artificial fertilization experiment, by exposing plants to air that was enriched to 8% carbon dioxide, and realized they grew much faster when exposed to a normal day–night light cycle.[12] He also showed that the green parts of plants take up and, importantly, also decompose a substance of the air while they assimilate water. In fact, he identified water as a key nutrient for plants. He also made sense of the complicated system that we outlined in Figure 5.1; plants use oxygen in respiration to provide energy and produce it when photosynthesizing. These two processes interfere in most of the early experiments; hence it was rather difficult to carefully separate them. There was one thing he was wrong about though; while he appreciated the important role of water for plants, he thought that the oxygen released under light conditions came from the carbon dioxide (fixed air). But this is for later…

Senebier and his illustrious predecessors were, of course, not aware of how in detail photosynthesis works and how carbon is captured in detail in the leaves of plants. Photosynthesis is

---

[12] Nash, L., 1957. It is interesting to note that the question how plants respond to increased concentrations of CO_2 is still very much a contentious issue see for instance Jiang, M., et al., 2020.

essentially composed of two steps. In step 1, light is harvested and water is split, and in step 2 the carbon from the carbon dioxide is fixed into sugars. It would take another 150 years before the detailed mechanisms involved into those two steps were elucidated. Kären Nickelsen, the historian of science, in her recent book 'Explaining photosynthesis, models of biochemical mechanisms 1840–1960',[13] describes how in the nineteenth century photosynthesis was considered as a one-step model, in which the key factors— light, carbon dioxide (or carbonic acid), water, and chlorophyll— in a living cell almost miraculously produced carbohydrates and oxygen. While it was known that several steps would be needed to arrive at the end products, how and in which order these steps would be executed was basically unknown.

Justus von Liebig in Germany was one of the founders of the field of organic chemistry and the first proponent of mineral fertilizer, and had a lifelong interest in improving agriculture. He not only was interested in photosynthesis, but also worked on the development of synthetic dyes, high explosives, artificial fibres, and plastics. He is mostly remembered for his 'law of the minimum' that the nutrient that is most scarce is the one limiting growth, and not the total amount of nutrients available. Justus von Liebig was not a person known for his humbleness. He tried to address two important questions related to photosynthesis: i) how the carbon dioxide was reduced and ii) how the carbon units that remained were put into much larger entities of molecules, the sugars.[14] The first question posed a serious problem since carbon dioxide was then known (as now, since this is the reason why it stays so long, unaltered in the atmosphere) as a very

[13] Nickelsen, K., 2015
[14] Nickelsen, K., 2015

stable, chemically almost inert, component—the reason Black called it 'fixed air'. Liebig thought that as a first step, from CO_2, H_2O, light, and alkaline bases, under the influence of a vital force, oxalic acid was produced ($C_2H_2O_4$). This would then be transformed into citric, malic, or tartaric acid, in the process releasing, as in the first step, oxygen, O_2. These organic acids all have more carbon atoms than oxalic acid, so this would address the second question posed. The oxygen would still come from the carbon dioxide, as in Senebier's scheme. Another scheme was proposed by Baeyer, the chemist who was awarded the Nobel Prize in 1905 for his discovery of the synthesis of the plant dye, indigo. He postulated that formaldehyde (COH_2) was formed as an intermediate step after which glycerol would lead to the sugars. This is in itself an interesting hypothesis since the general way to describe carbohydrates is $(CH_2O)_n$, indeed a multiple of formaldehyde. The difference between having formaldehyde as an intermediate step or organic acids such as citric and malic acid sparked a lively discussion over a few decades, particularly in Germany, with the balance of favour tilting towards the formaldehyde hypothesis among scientists.[15] However, everybody still thought the oxygen released during photosynthesis came in some way from splitting CO_2 (see also Figure 5.4).

The Dutch born microbiologist, Kees van Niel, managed to determine that the oxygen in photosynthesis comes from splitting the water molecule rather than the carbon dioxide molecule. He moved to the US in 1928 after he had worked at Delft University and was offered a job by Baas Becking[16] who was to set up

---

[15] Nickelsen, K., 2015.
[16] Baas Becking is known as one of the founders of the field of geobiology, and he popularized the statement: 'Everything is everywhere, but the environment selects'. The

Timeline

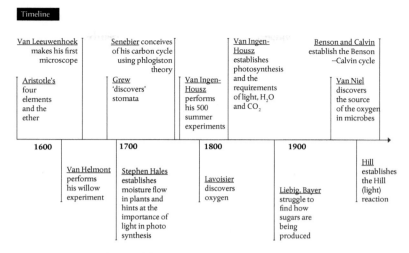

Figure 5.4 Timelines of the main events described in this chapter.

a new marine laboratory at Stanford University. During his stay at Stanford, Van Niel appears to have developed a lifelong liking for purple bacteria and in the USA he started working again with cyanobacteria. These were the subject of large confusion at the time: it was unclear whether their metabolic system was chemosynthetic, photosynthetic, none of these, or both. After six years of work van Niel concluded: '*photosynthesis can be considered as a light-dependent reaction in which different substances, specific for different kinds of photosynthetic organisms, serve as H-donors for the reduction of CO_2. This could be expressed by the generalized equation* $2H_2A + CO_2 \xrightarrow{\text{light}} 2A + H_2O + (CH_2O)$ *from which some important consequences could be deduced. First, the equation implies that photosynthesizing plants use H_2O as the exclusive H-donor and hence that the O_2*

statement implies that microbes in particular can be found everywhere, but that the specific conditions of the environment such as resource availability ultimately determine which ones will bloom.

they produce should be derived entirely from the de-hydrogenation of this substrate and not in whole or in part from CO_2.'[17] If we substitute H_2O for H_2A it is obvious that we have a way to perform photosynthesis; if we substitute H_2S for H_2A we have a way that another group of purple bacteria can grow. In short, there is unity in nature across plants and bacteria in the way they use light and carbon dioxide to grow! And the oxygen bubbles that Priestley and van Ingen-Housz saw came from the water rather than the carbonic acid. I find it hard to overstate the importance of this general formula, even though the precise mechanisms that were underlying this general formula still remained elusive.

Robin Hill took the important further step by isolating the main parts of the cell that were assumed to play a role in photosynthesis, the chloroplasts. These small organelles are the parts of the cell that execute the photosynthesis process and are easily visible through the existence of large amounts of chlorophyll. Chlorophyll is the light-absorbing pigment in photosynthesis. It is green because it mainly absorbs red and blue light. They are thought to originally have been single photosynthetic bacteria that were somehow incorporated into eukaryotic cells (cells containing a nucleus). Hill managed to isolate them by straining through glass wool. He was then able to show that a similar reaction as van Niel had found occurred that liberated oxygen. Importantly, *'The most suggestive view is to regard the chloroplast as containing a mechanism, the activity of which can be measured apart from the living cell, which under illumination simultaneously evolves oxygen and reduces some unknown substance which is not carbon dioxide.'[18]* By 1940 the situation was that the photochemical part of photosynthesis

[17] van Niel, C., 1967
[18] Hill, R. & Hopkins, F., 1939

was beginning to be understood and the release of oxygen known to be independent of carbon dioxide. This was a major step forward and the reaction expressing this $2A + 2H_2O \rightarrow 2AH_2+O_2$ is now known as the Hill reaction. The A stands for an artificial hydrogen acceptor. Using dark and light separately he was able to show that no oxygen was released in dark conditions. At least one part of the puzzle was becoming clear. The next big question lurking around was still how plants were able to produce sugars from CO_2, the Hill reaction showing how the energy could be generated, but not how that production process of sugars worked.

The sugar production process takes place in the Calvin cycle, the mechanism of which was elucidated in the 1950s[19] using unstable isotopes of carbon. Their properties make them an excellent tracer that can check where molecules containing carbon are produced. Isotopes are different variations of the same element, each with a different number of neutrons, causing a difference in the mass of the atom. There are stable and unstable isotopes: stable isotopes keep their mass and do not change; unstable isotopes lose their mass, primarily by radioactivity. The most common carbon isotope is ^{12}C, making up 98.9% of the total carbon—the stable isotope ^{13}C comprises 1.1%, while the unstable isotope ^{14}C makes up a minor $1.2 \times 10^{-10}\%$. ^{14}C is produced naturally in the atmosphere by cosmic-ray produced neutrons converting nitrogen into the ^{14}C isotope.[20] It falls back to nitrogen by emitting radiation, a so-called β-particle. The unstable ^{14}C isotope, also called radiocarbon, has a half-life of 5700 ± 30 years. This means

---

[19] Nash, L., 1957. Sharkey, T., 2019, and Nickelsen, K., 2015, Chapter 6, and Nickelsen, K., 2012, for more extensive and detailed histories of events.

[20] It is also produced in atomic explosions, hence since 1947 an increase in ^{14}C in the atmosphere has been observed that stopped after new tests were halted. This so called 'bomb radiocarbon' plays an important role in studying biogeochemistry of carbon, both on land and in the ocean.

that after 5700 years, half of the initial ^{14}C in the original reservoir remains; a quarter after 11,400 years and one eighth of the initial amount after 17,100 years. This makes radiocarbon, next to its use as a tracer, also an excellent isotope for determining the date of organic material. The dating technique using ^{14}C was developed by Walter Libby for which he was awarded the Nobel Prize in 1960. In a series of papers entitled 'The path of carbon in photosynthesis' (in total comprising about twenty-odd papers) they developed their theory. In the first report Calvin and his co-worker Benson showed in 1947 what happened when they held algae in light surrounded with N_2 to keep out CO_2 and then switched off the light and quickly added $^{14}CO_2$ (the notation here implies that the carbon in the carbon dioxide molecule is the heavy carbon isotope). They found a small amount of the labelled carbon (^{14}C) in sugars. This provided strong evidence at least some of the sugars were made after the light was turned off. *This paper was important because it demonstrated "that the reduction of CO_2 to sugars and the intermediates in that reduction does not involve the primary photochemical step itself" and that photosynthetic carbon metabolism "cannot be a simple reversal of the respiratory system of reactions".*[21] In the twenty-first paper of the series they came up with what is still considered the correct mechanism of how plants use CO_2 to produce sugars. They convincingly showed how CO_2 was carboxylated in a cyclic process. The original drawing of the Calvin cycle is given in Figure 5.5. This original drawing is both, in a scientific and aesthetic way, more appealing than the slick representations of the Calvin cycle one can find these days in textbooks and on the internet. In its bare essence, it somehow not only conveys the complexity of the pathways but also hints strongly at the difficult process that led to its discovery.

[21] Sharkey, T., 2019

The Calvin cycle takes up the CO_2 (left middle part) by a ribulose molecule (ribulose-1,5-phosphate, containing 5 C atoms), forming an intermediate, unstable β-keto acid (with 6 C atoms) that subsequently splits into two C3 molecules of glyceric acid (PGA or 3-phosphoglycerate). This is transformed into another C3 molecule (1,3-bisphosphoglycerate) that either is used to form sugars (hexose, 6 C atoms), polysaccharides (multi C atoms) in the top right of the cycle), or is being regenerated to form another C5 ribulose molecule that can act again as acceptor. The different elements of this scheme all have implications for how the ^{14}C used in the experiments would end up in the intermediate and final components and it took Calvin's group several years of painstaking analysis to complete the picture. These analyses were published in the preceding 20 papers. Let us also not forget that at the top left, there is the light-using process that produces oxygen and generates the energy (E) to convert the glyceric acid. The way living systems store energy is by molecules called ADP (adenosine diphosphate) and ATP (adenosine triphosphate). Converting ATP into ADP is similar to oxidation and generates energy. In the light reaction of photosynthesis at the top left, ATP is formed. ATP is used in the carboxylation step and in the regeneration towards the ribulose acceptor. While this is a necessarily short description, and some elements such as the cell's reducing agent NADPH (H in the figure) are missing, the Calvin cycle essentially performs three things: it fixes carbon (1), it then reduces it (2) and it then regenerates ribulose to begin again (3). In the whole process, sugars are being formed that can be used in the plant's or microbe's metabolism (top right exit Figure 5.5). It would be hard to think of a cleverer system!

In addition, Figure 5.5 leads us to the probably most important enzyme in the carbon cycle and Earth system. There are several

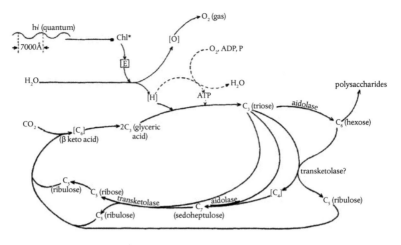

Figure 5.5 The Calvin cycle. The original caption in the paper reads *'Proposed cycle for carbon reduction in photosynthesis. Heavy lines indicate transformations of carbon compounds, light lines the path of conversion of radiant energy to chemical energy and the subsequent use of this energy stored momentarily in some compound (E), to form a reducing agent [H] and oxygen from water'.*
From Bassham, J., et al., 1954

names in the figure ending in '-ase', such as transketolase. This is the enzyme that catalyses and performs the reaction of making a ribulose (C5) from a hexose (C6) or vice versa. The important enzyme that performs the carboxylation step is called Rubisco, or technically Ribulose-1,5-bisphosphate carboxylase/oxygenase. Rubisco is sometimes called the 'schizophrenic' enzyme, the reason being that it can perform both the carboxylase activity, but also an oxygenase activity when it binds oxygen rather than carbon dioxide. This process is called photorespiration, as it inevitably is a consequence of the production and hence availability of oxygen in the first light step of photosynthesis. Rubisco is probably the most abundant enzyme on Earth, and arguably one of the most important. We estimate that about 120 Pg C (1 peta gram is 10^{15} gram) is annually sequestered by photosynthesis in the form of CO_2. This makes a whopping 16.42×10^{38} molecules of CO_2 being sequestered every year (10^{38} is the scientific notation for expressing a very large number, in this case a one with 38 zeros behind it!). Rubisco is the single enzyme that achieves all this. The current understanding is that the first photosynthetic bacteria appeared between 3 and 2.7 billion years ago (Chapter 6). Photosynthesis is thus not a recent evolutionary invention. In fact, the Rubisco-based system as we now know it has changed remarkably little from those we think were around for the first time on the planet some 3 billion years ago.

In Figure 5.6 our current understanding of the plant-related part of the carbon cycle is shown (compare with Figure 5.3 of Senebier's phlogiston scheme). CO_2 is taken up by the green parts of the plants with the help of the Rubisco enzyme, energy provided by the absorption of light, releasing free oxygen in the process. It is then converted into sugars through the Calvin cycle and used in the plant's metabolism to further growth and converted into other products such as proteins. This metabolic process, called respiration, requires oxygen and moves the carbon of the plants back into the atmosphere. Fallen leaves and other dead material are stored in the soil or decomposed by a variety of microbes in the soil, and ultimately also converted back to CO_2 in the atmosphere. In the long run, photosynthesis should equal the respiration fluxes, but in practice various disturbances, such as forest fires, droughts, increases in the availability of nutrients or other environmental causes can make the balance temporarily shift towards a negative or positive value.

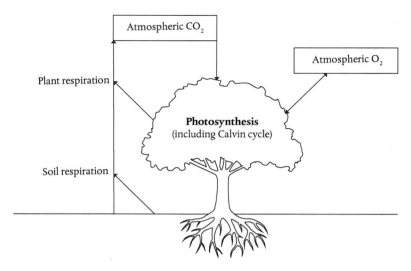

Figure 5.6 Our current understanding of the interaction of photosynthesis and respiration in the carbon cycle. Compare also with Figure 5.3.

We are now finally able to build a picture of the role of CO_2 in the Earth's terrestrial ecosystems. Plants take up CO_2 through photosynthesis and convert it into sugars that are used to further build plant material. This plant material dies off, the carbon being oxidized (respired) back into the atmosphere as carbon dioxide. It took our cast of selected scientists starting from Jean Baptista van Helmont to Melvin Calvin more than 200 years to unravel the story.

CHAPTER 6

CARBON DIOXIDE IN THE GEOLOGICAL PAST

W e trace CO_2 concentrations through geological time, some 4.5 billion years in total, from the start of the formation of the Earth to the beginning of the last ice age. After the core and the crust were formed— and the Moon extracted—the Earth started to develop an atmosphere. An atmosphere that consisted of the exhaust gases from the interior; and an atmosphere that may have been blasted away quite a few times in its early history by large meteorites. This atmosphere most probably consisted of water vapour, nitrogen, and carbon dioxide. It was unlikely to have been a pleasant place, hot and continuously bombarded by meteorites. It is thus with some reason called the Hadean, after the Greek god of death, Hades. The initial pressure of CO_2 may have been 10–100 times the current atmospheric pressure; however, 4 billion years ago, CO_2 probably made up 70% of the atmosphere. But how was it possible that when the Sun was less radiant in its early stages, the Earth did not freeze over? Yes, most likely the greenhouse gases CH_4 and CO_2 worked to keep temperatures above freezing.

During the next geological period, the Archean, covering only 1.3 billion years, life started to appear. From palaeosols and understanding the weathering process, it is possible to back-calculate the atmospheric CO_2 concentration over the Archean. This would drop to values around 10–20 times *the current one. But then, life developed, first as ancestors of bacteria and archea that produced biomass without fully using photosynthesis. The ability of cyanobacteria to photosynthesize meant that slowly the atmosphere changed because oxygen was produced. This period is called the Proterozoic and lasts from 2.5 billion years ago (Gyr) until 541 million years ago (Myr). By that time the atmospheric concentration of CO_2 had gone down to roughly 3–5 times our current level, around 1000–2000 ppm. The availability of oxygen and ozone that provided a shield against ultraviolet, now generated an explosion of new lifeforms. During this period Earth experienced a large drop in atmospheric CO_2 during the Devonian (419–359 Myr) and the Carboniferous (359–299 Myr) due to the appearance of land plants. These plants would decompose but the remaining biomass would be buried in soil and ocean and transformed into our future fossil fuels.*

From this Hothouse Earth in the Eocene down to about 2.5 Myr when the Pleistocene started, the CO_2 concentration would continue to decline, with occasionally large spikes, likely due to periods of increased volcanic eruptions. These spikes, such as the Palaeocene–Eocene Thermal Maximum (PETM) at 55.8 Myr provide paleoclimatologists with analogues for current and future climate change. During the PETM the temperature rose 5 °C and the oceans acidified, with major extinctions as a result. Around 34 Myr, at the Eocene–Oligocene transition, small changes in the balance between the sources of CO_2 through volcanic eruptions and sinks through weathering had brought down the atmospheric concentration to around 400–900 ppm. Earth was moving from a hothouse into an icehouse, and ice sheets had started to appear.

There might, after all, be some truth in Genesis 1, the Bible verse about the creation of the world: 'The earth was without form and void, and darkness was over the face of the deep'. 4.6 billion years (Gyr) ago when our planetary system formed, the Earth was one of the few planetary embryos that were shaped through gravitational collapse or accretion out of the solar nebula, a sort of left-over debris after the Sun had formed. It took Earth an estimated 10–100 million years to acquire its shape. During that time Earth was under a continuous bombardment of left-over rocks and dust (meteorites) within the nebula. Generally, the birthyear of Earth as a separate planet is set at 4.45 Gyr.

Sometime after the first thirty to fifty million years, the Moon was formed. At least, that is our current understanding. The most favoured theory for the formation of the Moon is that there was a catastrophic collision between an early planet, the size of Mars, and a proto-Earth which would have had 90% of its current mass. This planet, sometimes called 'Theia' after the mother of Selene, the goddess of the Moon, struck our proto-Earth, ripping off a piece that would form the Moon. Earth at that time was not a pleasant place to be and a far cry from the habitable planet we know now. The impact would not only have caused the mantle to melt and give rise to temperatures more than 5000°C but would also probably have—at least partially—destroyed Earth's first atmosphere.[1] This atmosphere would have consisted of volatile compounds from vaporized rock material. However, it is quite possible that when the other planet collided with it, Earth's early atmosphere had already been lost by previous large collisions. A lot is uncertain about the very early Earth.

---

[1] Zahnle, K., et al., 2007

After the Earth had passed its initial accretion phase and the Moon-forming impact, an inner and outer core, a mantle, lithosphere, and crust were being formed. These are the ingredients of the Earth as we know it now. Exhaust gases from these components, primarily from the mantle and crust, then started to populate the primordial atmosphere. Some of these gases may also have originated from the impacts of large meteorites that would have been vaporized by the impact of the collision with Earth. Meteorites that hit the Earth during its early formation carried frozen water and nitrogen, elements that helped to make Earth a habitable place in the future. The main volatile gases forming this early atmosphere were water vapour, carbon dioxide, and nitrogen together with chloric acid and hydrogen sulfide. The latter two gases were very likely rained out leaving an atmosphere that was composed mainly of water vapour, nitrogen, and carbon dioxide.

We have very little information about the early Earth as virtually every piece of rock that could have existed at that time has by now been recycled. However, tiny (smaller than 0.5 mm) crystals of the mineral zircon have resisted decomposition and these are regarded as the oldest survivors, estimated to be 4.4 Gyr old. These extremely durable crystals indicate that they were formed from a magmatic source, like the formation of the current continental crust. Luckily, the specific conditions in which they were formed left a trace in the isotopes of oxygen and this allows us to infer that they were likely formed in interaction with a liquid hydrosphere. In other words, their presence and isotopic composition shows that a continental crust was being formed at that time in and around oceans, or that at least that liquid water may have been present. This does rather call into question the often repeated 'hell-like' conditions of the early Earth, with soaring temperatures up to 2000–3000 °C. This period (4.5–3.8 Gyr) is often referred to as the Hadean after the Greek god of death, Hades.[2]

The presence of water when the zircon crystals were being formed is one of the indications that the Earth could have been much cooler—at a low temperature around 500 °C. In fact, one can have liquid water at those temperatures when the pressure is as high as 100 times the current atmospheric pressure, which could have been the case. It is one of the conundrums of the Hadean period that although we are free to develop theories, hardly any material exists to confront those theories with experimental data. Thus, while the existence of such a cool period is not really debated anymore, the length and exact timing, and actual temperatures are. During this period, from 4.4 to 4 Gyr, atmospheric partial pressure of CO_2 may have dropped substantially from a whopping initial value of around 10–100 times the current atmospheric pressure (101.3 kPa!) and a 70% contribution by CO_2, to values around 100 kPa after what is called the Late Heavy Bombardment. This period was one in which the Earth was, once again, frequently bombarded by meteorites. We know some details of this period largely thanks to the landings of the Apollo crews on the Moon. They brought back rocks of that age from the lunar surface where they had been lying, virtually undisturbed, since 3.9 Gyr. During this period, it is quite likely that the Earth's oceans would have completely evaporated, and

---

[2] See https://www.bgs.ac.uk/discovering-geology/fossils-and-geological-time/geo-logical-timechart/ for a geological timescale with all the relevant periods.

its atmosphere lost several times during the impacts of the more massive meteorites.[3]

After the first 800 million years or so, the Earth was beginning to find something of a rhythm: a solid crust, beginnings of tectonic movement, and, importantly, the start of a carbon cycle. It would have taken anywhere between 10 and 100 million years to take up most of the atmospheric CO_2 at the enormously high initial partial pressures in the Earth's developing geological systems. Once most of this CO_2 was taken up and subducted into the mantle, a balance would have been slowly established between rock sinks such as those produced by weathering at the surface (Chapter 4) and reaction of CO_2 with the oceanic crust to form carbonates.

Unfortunately, there are no sedimentary rocks of that time that would allow us to reconstruct the atmospheric composition. Sedimentary rocks are the bread and butter of geologists as they allow them to determine geological processes that are correlated in time. Individual layers of sediments are piled upon each other in the ocean and when compacted under their collective weight, they can form rocks—as Hutton had been one of the first to realize. Within these layered structures geologists can discover the chronology of the sedimentation and thus reconstruct the environment during sedimentation. The oldest such records we have are metamorphosed (changed by heat and pressure, see Chapter 5) sedimentary rocks in Greenland, called the Isua rocks. These rocks contain finely laminated layers of slate. They are dated at 3.8 Gyr and are generally known as the oldest Archean rocks. The Archean is the next geological period, taking

---

[3] Marchi, S., et al., 2014

us from the Late Heavy Bombardment up to 2.5 Gyr, covering a period of 1.3 billion years (1,300,000,000 years). Life first appeared on Earth during the Archean—although the precise timing is hotly debated. Some put this start at the beginning of the Archean because of small grains of carbon in the Isua rock that show depletion of the heavy carbon isotope (^{13}C) that is possibly related to CO_2 fixation. This debate about the origin of life is a key question for humans and many books and scientific papers have been written about it (e.g.[4]). For us, the interesting question is: How much CO_2 was in the atmosphere at that time?

There is, however, another important problem that needs to be solved first. This is called the faint young Sun paradox, and its solution brings us a whole lot closer to understanding the role of CO_2 in the climates of the past. The faint young Sun paradox was first raised by the scientist and writer Carl Sagan and his co-worker George Mullen.[5] Sagan was a scientist with a wide range of interests: from astronomy to astrobiology to planetary science. He is primarily known for his work as a science communicator and was for instance involved in developing the concept of the nuclear winter that he thought could occur after nuclear explosions. He argued that these explosions would bring so much aerosol up into the stratosphere that the Earth would cool for a prolonged time after. He was also one of the people who incorporated the golden plate on Voyager I and II, the first spacecraft to explore our outer planets Jupiter, Saturn, and Uranus, and after that the interstellar space outside our solar planetary system. The golden plate played sounds from Earth that, so it was hoped, could in principle be understood by intelligent life.

[4] E.g. Lenton, T. & Watson, A., 2011.
[5] Sagan, C. & Mullen, G., 1972

When the Sun formed at 4.6 Gyr it did not shine as brightly as it does today. It is estimated that at about 4.2 Gyr the Sun emitted radiation at roughly 70% of its current strength. This is important, as before the Sun arrived at a fairly stable radiation emission level, the available water on Earth could be either in a gaseous form when the luminosity was temporarily higher, or frozen when the luminosity was low. With only 70% of the incoming radiation the question is how the Earth avoided being completely frozen over. How could Earth have created such circumstances that would make water available in its three forms: gas, liquid, and ice? And perhaps more importantly, how could Earth have created the conditions for life to have evolved during the Archean?

Without an atmosphere the surface temperature of the current Earth would be −18 °C. During the Archean, with the Sun only at 70% of its current strength, the temperature without an atmosphere would be a rather freezing −40 °C. Our current greenhouse effect adds about 33 °C to make a comfortable 15 °C possible (−18+33 = 15), but in the Archean adding this greenhouse effect would only lead to −7 °C, so still freezing water. We do know, however, from our zircon crystals and the plain fact that from 3.8 Gyr we have sedimentary rocks that were originally deposited in ocean water, that liquid water was available at that time, and probably earlier.

Sagan and Mullen correctly worked out that the surface temperature would need to be reinforced by some greenhouse effect to achieve temperatures that keep water liquid. They also realized that levels of atmospheric oxygen were very low before 2.4 Gyr: in chemical terms he assumed that the atmosphere was in a non-oxygenized (reduced) form, probably also containing hydrogen. Thus, they suggested that enhanced concentrations of these reduced greenhouse gases, specifically ammonia (NH_3) and methane (CH_4), could have kept the early Earth warm. However, quite soon after publication these ideas were criticized and several other solutions to the problem were put forward. It was pointed out that ammonia would be chemically unstable and would thus not have remained in the atmosphere in sufficient amounts. The impact of water vapour as a greenhouse gas was also mentioned, but since water condenses in clouds this cannot really serve as an explanation for the required greenhouse effect. Others suggested CO_2 and this seems at present the most likely solution. Ideally, we would have proxies that would allow us to precisely reconstruct the CO_2 concentrations in the Archean. Proxies are biogeochemical indicators that allow us to reconstruct a particular variable in the past.

One of those proxies that help us to reconstruct CO_2 levels in the Archean are old palaeosols, fossil soils that have been exposed to the Archean atmospheres and that have turned into rocks. They bear information about the composition of this early atmosphere: when CO_2 dissolves in water it produces acid (see also Chapter 4). Acidic water would have dissolved key components of the original rocks and in doing so produced new soil. By back-calculating how much is left in the fossil soil, we can come up with an estimate of the then dominant concentration of CO_2 in the atmosphere. This yields estimates that are easily 20–50 times higher than the current level for 2.8 Gyr (Figure 6.2). The reason why we can calculate this at all is, of course, precisely what makes CO_2 such an important molecule in the Earth system. The interaction of CO_2 with the rock surface that we discussed earlier is one of the key processes for removing CO_2 from the air. Once again, the tectonic cycle is key to the recycling of CO_2 at these large timescales. Figure 6.1

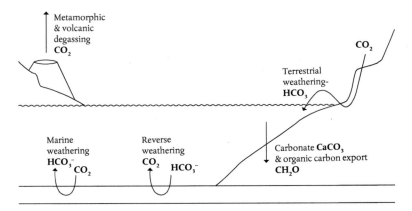

Figure 6.1 Simplified diagram showing the geological (inorganic) carbon cycle that determines at long timescales the atmospheric CO_2 concentration. Note the importance of marine weathering next to classic continental weathering (as in Figure 4.4).
Redrawn after Isson, T., et al., 2020

shows the important processes of the inorganic carbon cycle in a simple diagram. While the palaeosols we mentioned earlier are the remnants of the continental weathering, sediments allow us to investigate ocean processes and the chemical conditions in which they were formed. In the long term the processes that produce CO_2 (sources) must balance those that consume CO_2 (sinks).

In chemical notation this reads as $XSiO_3 + CO_2 + H_2O \leftrightarrow XCO_3 + SiO_2 + H_2O$ where X is a cation (an ion with a positive charge, the name coming from the fact that in a solution it would move to the negatively charged cathode, as opposed to negatively charged anions that would move to the anode). SiO_2, silicon dioxide is also known as chert. This is the familiar Urey reaction we discussed in Chapter 4. Going the reverse way, it produces CO_2. In the Archean it is likely that there was more seafloor production and seafloor weathering was probably as important as continental weathering, with the alkalinity of the ocean providing a key control.

So where does all this leave us in the search for a reliable estimate of the amount of CO_2 in the Archean atmosphere? It is likely that around the start of the Archean, around 4 Gyr, concentration levels would have been 100 to 100000 times the present one (PAL, Present Atmospheric Level). Over the course of the Archean the CO_2 concentration would fall as the Earth's geological cycle came to grips with it, and CO_2 levels would drop to only 10–20 times the present one (see Figure 6.2). The initial high concentration levels of CO_2 could have countered the reduced solar input in the Archean. It may have had some help from its sibling greenhouse gas CH_4 which is about 25 times as strong as CO_2 in its effect on climate.

Now it is time to introduce the role of life in the early atmospheric composition—we have avoided it for too long. The earliest appearance of life is still subject to debate and varies from 2.7 Gyr to 3.8 Gyr; the latter based on specific structures found in some of the most ancient sedimentary rocks (the Isua group in Greenland). The first appearance of life that could affect the CO_2 in the atmosphere might have been the ancestors of bacteria and archaea. Already some of the earliest microbes may have been able to use inorganic compounds (e.g. H_2) as an energy source and draw CO_2 out of the air to convert it into biomass (using one of the earliest evolved pathways for carbon fixation) or into methane via an energy-conserving metabolism referred to as methanogenesis. This could have been an important mechanism to produce biomass early on in Earth's development. Others were fermentative or grew by anaerobic respiration using organic compounds as the source of energy. Oxygen was probably poisonous to most of the early microbes on Earth and only became more abundant on Earth upon the evolution of Kees van Niel's friends, the cyanobacteria, which produce molecular oxygen from water using sunlight as the energy source. Indeed, the availability of oxygen drastically changed the Earth's chemistry. For instance, iron would be freely available in a non-oxygenated world. With the occurrence of oxygen, the ferrous (Fe^{++}) form would disappear from the ocean, making iron in the ocean a scarce material. This shortage of iron would have often constrained the growth of microbes and the later evolving eukaryotes including algae, as it does in the current ocean. On land however, iron in its oxidized form (ferric, Fe^{+++}) would become ubiquitous, as shown by palaeosols.

The oxygenation of the Earth marks the start of the next big geological eon, the Proterozoic. Together with the Hadean and Archean, the Proterozic completes the Precambrian, the period before the start of the Cambrian at 541 Myr (yes, we are getting closer to our time, and talking millions of years now!). The Proterozoic eon starts at around 2.5 Gyr, so comprises another 2 billion years. The big push for oxygenation of the Earth came at around 2.4 Gyr. This is arguably the single most important biogeochemical revolution on our planet, with massive implications. It would, for instance, have removed much of the haze that is assumed to have characterized the atmosphere from its early stages and has provided the reddish, dark brown colour that one finds in artistic impressions of that time. The presence of oxygen in larger amounts would also have helped life to explore new niches because ozone (O_3), formed from O_2, would have provided increased protection against harmful UV radiation. It is worth noting that the whole process of oxygenation of the planet took probably hundreds of millions of years and early forms of life likely produced and consumed small amounts of oxygen. But by 2.5–2.3 Gyr the Earth's atmosphere contained roughly 1% of the current amount of oxygen.

Another impact, that relates more directly to our subject, is that an atmosphere with oxygen would not be able to hold many chemically reduced components, such as NH_3 and CH_4. In fact, the rise of oxygen and subsequent oxidation of CH_4 to CO_2— recall that CH_4 is a much stronger greenhouse gas (GHG)— may have pushed the Earth into one of its snowball states, where most of the surface and oceans would be frozen over. The evidence for this comes from several glacial deposit remains in various places over the Earth, including those that reflect ice at the Equator. It is still something of an open question whether the Earth was com-pletely covered or whether there were still liquid waters around, the latter option appropriately called a slushball Earth. Interest-ingly, it could have been our geological carbon cycle that provided the escape option from the snowball Earth, as decreasing temper-atures decrease weathering and thus may have halted the further removal of CO_2 from the atmosphere.

Figure 6.2 shows the trend of atmospheric CO_2 based on an original plot by James Kasting, one of the scientists who have published widely on the early Earth atmosphere and its implica-tion for life.[6] During the snowball periods such as the Huronian glaciation, the concentration of CO_2 dropped a little more but returned back relatively quickly (geologically speaking).

By the end of the Proterozoic, the now plentiful supply of oxy-gen allowed eukaryotic life including animals, plants, and fungi

---

[6] Kasting, J., 1993

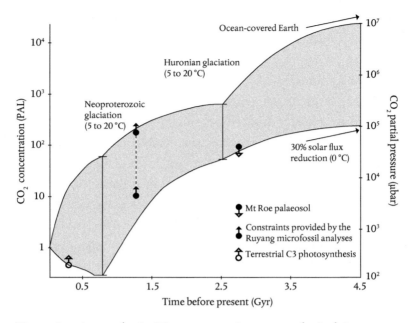

Figure 6.2 Atmospheric CO_2 concentration over geological time expressed as multiples of PAL (present atmospheric level) and corresponding partial pressure levels of CO_2 in the Archean (4–2.5 Gyr) and Proterozoic (2.5–0.5 Gyr). Note, 10^6 μbar is 100 kPa, and 1 μbar corresponds roughly to 1 ppm. The shaded area represents an uncertainty band based on a model that calculates the assumed CO_2 level to counteract the faint young Sun low radiation. Several constraints from reconstructions such as from palaeosols and microfossils are also indicated.
From Kaufman, A. & Xiao, S., 2003

to diversify. The most famous relics of this are a series of fossils in the Burgess shales (in British Columbia, Canada) that show an incredible diversity of prototype species, from which only a few remained and passed through the evolutionary selection process. One of the weirdest is no doubt *Opabinia*, a five-eyed, soft-bodied animal species with a fan-shaped tail and a hoover-like extension to its mouth with which it could scavenge soft small food from the seafloor. It is now, one could say, unfortunately, extinct. This occurrence of such a large diversity of animals is what is known as the Cambrian explosion. For our purpose it is important to realize that animals require oxygen to burn with organic compounds such as sugars, proteins, and lipids to conserve energy: similar to many microbes, they are organoheterotrophs, ultimately dependent on autotrophic organisms, that can convert inorganic carbon to biomass using chemical or light energy.

When the Proterozic ended, the Phanerozoic, our current geological eon started. At the start of the Phanerozoic, CO_2 levels in the atmosphere were roughly 3–5 times as large as our current level, say 1000–2000 ppm. It is from this period onwards that, almost for the first time, we can reconstruct somewhat reliable CO_2 levels from various sources (proxies). Five such proxies are available to determine the Phanerozoic CO_2 record: stomata; phytoplankton and liverworts; palaeosols (we have encountered these before); boron; and nahcolite, a rare sodium carbonate mineral. Stomata are the tiny holes that were discovered by Nehemia Grew and Marcello Malpighi (Chapter 5) through which CO_2 is taken up and water released by plants.

Stomata are finely tuned to the atmosphere as approximately 200% and 16% of the total content of atmospheric water vapour and CO_2 are cycled through stomata each year. Calculating the occurrence or density of stomata on leaves allows us to reconstruct the CO_2 levels occurring at the time the stomates were fossilized. Since plants occurred from about 420 Myr, this is of course as far back as we can go by counting stomata to reconstruct CO_2. The oldest plant containing fossil stomata is the now extinct species *Cooksonia* from the lower Devonian period in the

Paleozoic.[7] Liverworts are small plants, whose first occurrence was established in the Ordovician (470 Myr). They belong to the bryophytes, as do mosses, and importantly have no stomata. When CO_2 is abundant they have little preference for the lighter isotope ^{12}C above the heavier one ^{13}C. The reverse holds when CO_2 declines: then the difference between the two isotopes becomes larger and more ^{13}C is incorporated, and in this way the $\delta^{13}C$ of the organic material can be used to reconstruct atmospheric CO_2. In the sea, phytoplankton works in a similar way, but one also needs to know the ambient sea water value to be able to reconstruct the precise value for the atmosphere. The use of boron relies on the fact that the relative proportions of the two major boron species in the ocean, $B(OH)_3$ and $B(OH)_4$, vary with pH. Because the isotopic composition of these two species differs, this difference in marine sediments can be used as a proxy for the pH of the sea water. From the pH, with additional assumptions about the total alkalinity, atmospheric CO_2 can then be inferred. Our final proxy, nahcolite, is a rare sodium carbonate mineral that only precipitates when the CO_2 level is high (1330 ppm). When it is found it is thus indicative of a minimum level of atmospheric CO_2. Each of these proxies has its problems. When, for instance, the atmospheric CO_2 concentration is beyond 1000 ppm, stomata become quite insensitive to further increases. The biological proxies such as liverworts and phytoplankton suffer from the possibility that growth rates of the organisms may also impact the isotopic fractionation of carbon and thus give a false picture of the dependence on only atmospheric CO_2.

---

[7] Berry, J., Beerling, D., & Franks, P., 2010

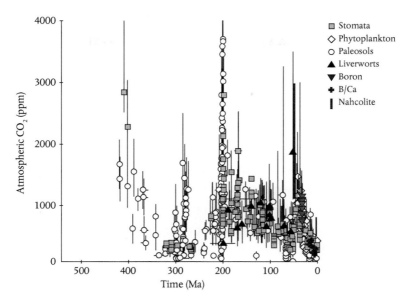

Figure 6.3 History of atmospheric CO_2 during the Phanerozoic from proxies based on several individual CO_2 estimates.
From Royer, D., 2014

Despite all these problems, scientists have developed a widely accepted reconstruction of the decrease of CO_2 during the Phanerozoic (Figure 6.3). The figure shows a large drop in atmospheric CO_2 during the Devonian (419–359 Myr) and Carboniferous (359–299 Myr) periods. This is the time when land plants first appear in the geological record. By the mid-Devonian, trees were probably up to 3 m high, but reaching to over 8 m by 385 Myr, spanning a period of 7 Myr only. The first recognizable tree is *Archaeopteris*, which had a stem, that divided into branches and ultimately leaves. It could have reached heights of 40 m and looked a bit like a modern-day conifer.

Remember the chemical equation of the organic carbon cycle: $CH_2O + O_2 \leftrightarrow CO_2 + H_2O$? In essence this tells us that if we burn (oxidize or respire) sugars, we produce carbon dioxide and water. However, the reverse also holds: CO_2 and H_2O can be used in photosynthesis to produce sugars. Hence, in the long term this does not yield either oxygen or carbon dioxide as the two processes need to balance. This is a classic zero-sum game in terms of atmospheric carbon dioxide and oxygen. What is not a zero-sum game, however, is when we take out carbon from the long-term carbon cycle by burying the carbon material as sediment. Robert Berner has termed this process net photosynthesis. Berner was a geologist with an interest in oxygen and CO_2 in the Phanerozoic who died in 2015. Net photosynthesis is photosynthesis minus respiration, but importantly, as represented by the burial of organic matter. He calls the reverse process, the oxidation of carbon *georespiration* and we might call his net photosynthesis in a similar vein *geophotosynthesis*. The chemical equations are the same, but the timescales on which they operate are fundamentally different.

So, when the first vascular plants arrived, they took up carbon in the form of carbon dioxide from the air and produced organic material. A lot of that would be respired back again into the atmosphere, but as we have noticed before, not all. We do after all owe our fossil fuel stocks (next chapter) to the fact that these organic materials became buried in sediments and started a conversion process under the pressure of the overlying rock layers to coal and oil. How can we calculate the fraction of organic material that gets buried in the sediments, the burial ratio? We make use once again of the isotopes of carbon. The Rubisco enzyme that forms part of the plants' Calvin cycle (Chapter 5) prefers the lighter variant of the carbon atom (^{12}C) above the heavier (^{13}C). In organic material that gets buried on land we therefore tend to find more of the lighter carbon and thus less of it in the atmosphere. The atmospheric carbon then equilibrates with the ocean waters, so the carbon taken up in the ocean by algae reflects the heavier carbon. All in all, then, the carbonates in the sediments that are high in the ratio of ^{13}C to ^{12}C are indicative of increased burial of organic material. Are you still with us? Prepare for the next surprise: the burial ratio has been remarkably constant during the last 3.5 Gyrs, at about 0.2. In other words, the carbon burial mechanism may not have been the key factor explaining the decrease in ambient CO_2 levels. But what then can explain this drop in the Devonian, from around 400 to 360 Myr?

Current theory favours the idea that plants played an important role, but rather more by triggering the geological cycle, than through their direct uptake of CO_2. Geology rules—particularly at long timescales! It is believed that they increased the weathering by forming soils out of the rocks by their roots. This would accelerate the uptake of CO_2 through the geological carbon cycle and decrease the atmospheric CO_2 content, provided there was no increase for instance in volcanic degassing. The data from palaeosols of the Devonian appears to back up this scenario. The ability of rocks to weather, their weatherability, depends on a lot of factors such as lithology (the type of rock), tectonic activity, the configuration of continents, and biological (root) and hydrological (availability of water) activity. What is known is that increased weatherability lowers the atmospheric CO_2 concentration, as in the long term the sources of CO_2 (volcanic outgassing) must balance their sinks (the weathering). It is important to highlight here that the atmosphere contains tiny amounts of carbon compared to the geological reservoirs. Any change in the atmospheric concentration at short time scales (<1 Myr) has the potential to quickly trigger runaway conditions in either a hothouse or an icehouse. The stability of the temperature record, however, suggests that Earth has generally maintained a nice balance between the sinks and sources of carbon. Thanks for this are due to the geological carbon cycle with its stabilizing feedbacks, such as terrestrial and marine silicate weathering and the reverse silicate weathering feedback.

However, there have been quite a few excursions during which temperatures and carbon dioxide levels at geological timescale quickly rose and declined. We now turn our attention to a few of those (Figure 6.4). These events might provide some useful insights as climate analogies of our current climate anomaly, as several geologists have noted. Much of the early Cenozoic (the last 65 million years of Earth's history) was characterized by noticeably high concentrations of greenhouse gases, as well as a much warmer mean global temperature and poles with no ice.[8] A quite abnormal (although one could ask what is normal and abnormal in geology) event took place around 55.8 Myr, and this has become something of a *cause célèbre* among geologists and palaeoclimatologists. During the Palaeocene–Eocene Thermal Maximum (PETM) (55.8 Myr) the planet suddenly warmed by 5 °C, and a series of deep-sea organisms became extinct, while the oceans rapidly acidified. Global temperatures rose within 20,000 years and remained at high levels for another 100,000 yrs. The short time in which this change took place and the subsequent recovery make it a favourite for studying the response of the Earth system to drastic changes in the carbon cycle. It is recognized by a sudden

---

[8] Zachos, J., Dickens, G., & Zeebe, R., 2008

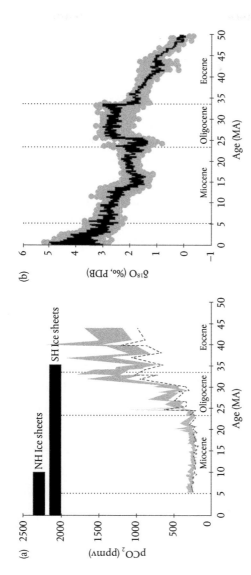

Figure 6.4 Reconstructed CO_2 concentrations (ppm) (a) and $\delta^{18}O$ (b), an indication for temperature (going from warm to cold) from the Eocene to the Miocene; bars indicate the existence of large ice sheets in the Southern and Northern Hemisphere.

From Pagani, M., et al., 2005

negative jump (a negative isotope excursion) in the $\delta^{13}C$ ratio of atmospheric CO_2, indicating the release of large quantities of light carbon (^{12}C). The most likely culprits for this injection are volcanic CO_2 and CH_4, most likely from ocean clathrates, but increased decomposition from terrestrial sources may have contributed. The methane would be very effective initially at enhancing the greenhouse effect, but also relatively quickly oxidized to CO_2. While the Eocene hyperthermals provide good case studies for the response of the climate system to changes in the carbon cycle; they are also pretty hard to study. For instance, getting good pCO_2 estimates from the proxies such as palaeosols and, importantly, the accurate age associated with them, still presents major problems. Hence it is often difficult to reconstruct the precise variation in atmospheric CO_2 during these events.

The most extreme of these hyperthermal cases is the Early Eocene Climatic Optimum (EECO), 53–51 Myr, when pCO_2 was high, maybe approaching 4000 ppm (see also Figure 6.3, about ten times our current value) and global temperature reached a long-term maximum (24–30 °C). Geologists speak, with good reason, of this period as a 'Hothouse Earth'. This period is almost the last of the high CO_2 concentration eras of the recent geological record. It is believed to have been caused by relatively high volcanic emissions during the period of 40–60 Myr and changes in weathering associated with the geological cycle. After the EECO the atmospheric concentration of CO_2 and temperatures generally went downhill, and Antarctic ice sheets start to appear, the first about 33.7 Myr. Arctic ice sheets are a more novel feature of our recent geological past—they appeared towards the end of the Miocene, about 10 Myr. Earth was on course to being an icehouse. But how?

While most of the information about environmental conditions in early geological history comes from analysis of rock samples, for more recent geological periods, particularly the Cenozoic era, ocean sediments provide the key information to the palaeoclimatologist. These, however, need to be retrieved from the ocean bottom, sometimes from great depths (more than 2000 m). The first systematic study of the global ocean and seafloor was performed by *HMS Challenger* (1871–1876). This largely involved collecting grab samples from the ocean floor. An important step forward in the recovery and investigation of deep-sea sediments was the invention of the gravity corer by German researchers. This allowed recovery of continuous sections of sediment, restricted to 1–2 m in length. However, given the slow sedimentation rate (typically 10 mm /1 kyr), such a core would contain information on thousands of years only. It was the Swedish marine geologist Börje Kullenberg who came up with a new device, the piston corer. This was a modification of the gravity corer and opened the way to retrieving much longer (up to 10 m) virtually intact cores and consequently the analysis of much longer time periods. The piston corer was first used during the Swedish *Albatross* deep sea expedition in 1946. In 1968 oceanographic research got a further boost when the US started the Deep Sea Drilling Project (DSDP).

The Deep Sea Drilling Project began coring in August 1968. Funding and direction was given by the National Science Foundation's Ocean Sediment Coring Program. Their mandate was to increase the knowledge of the Earth's development through an ambitious ocean sediment coring programme. The Prime Contract for the Project was executed in 1966 between NSF and the University of California Board of Regents. The Scripps Institution of Oceanography, an integral part of the UC system, was responsible for managing the Project. Global Marine Inc., through a subcontract with Scripps, provided the drilling vessel and crew. Major oceanographic institutions of the United States supported the drilling programme by contributing to the planning of the scientific objectives. This loose collaboration developed into a more formal organization, the 'Joint Oceanographic Institutions for Deep Earth Sampling (JOIDES)'. Prompted by the initial scientific and technical successes in the first seven years, the Project increased the scope of the coring programme to include even deeper penetrations into the ocean floor. Several foreign scientific institutions were now getting interested in becoming members of JOIDES, leading to the 'International Phase of Ocean Drilling'. From 1983 to 2003 this developed into the Ocean Drilling Program, with its own ship the *JOIDES Resolution*. The programme now became more science-driven with proposals being solicited and evaluated by the JOIDES advisory committee. The funding was largely US-driven with the National Science Foundation providing about two-thirds of the funding. From 2003 to 2013 the Integrated Ocean Drilling Program and later 2013 to 2023, now known as the International Ocean Discovery Program (no need to change the acronym IODP), aimed to provide the drilling infrastructure through several ships and specifically designed mission platforms for drilling.

One such mission was Expedition 302 Arctic Coring—it provides some insights into the trials and tribulations of this type of palaeoclimatological research. It was the first IODP mission-specific platform operation managed by the European Consortium for Ocean Research Drilling. The expedition took place in the late summer of 2004. The drill sites were exceptional: they were located on the Lomonosov Ridge, at a point only 250 km from the North Pole. Most previous drilling had taken place in tropical, subtropical, or temperate waters. This location provided the scientists and engineers with a huge logistical challenge because the drillship had to hold its position while surrounded by the moving sea ice of the Arctic Ocean. The *Vidar Viking* icebreaker was specially converted for the task to undertake the coring. Two additional icebreakers, the *Oden* and *Sovetskiy Soyuz* were to crush large ice-floes into small pieces allowing the *Vidar Viking* to maintain position. The primary goal of the expedition was the recovery of a sedimentary sequence more than 400 m thick, draping the crest of the Lomonosov Ridge in the central Arctic Ocean between 87 °N and 88 °N. From analysis of the cores the palaeoclimatologists hoped to determine the Cenozoic palaeo-environmental evolution in the central Arctic Ocean and the Arctic's role in the global development from the Palaeogene greenhouse to the Neogene icehouse. They drilled five holes at three sites into the Cenozoic sediment providing the first long record of Cenozoic sediments from the central Arctic Ocean.

The core recovery was not a total success: the average core recovery was only 68.4% and below 270 m at around 47 Myr, the recovery rate fell even further to 43%, implying that several geological episodes were missing from the sedimentary sequences.[9] Despite this setback, in the sediments of the retrieved cores scientists found very high concentrations of spores of the freshwater fern *Azolla*. The timing of this growth coincided approximately with the onset of a global shift towards heavier deep sea benthic foraminifera $\delta^{13}C$ values and an overall global cooling trend that

[9] Backman, J., et al., 2006

we have seen before. Could these large quantities of *Azolla*, float-ing in a brackish or fresh water layer on top of the salty ocean have helped to bring down the atmospheric CO_2 concentrations? *Azolla* is a freshwater plant, and one of the fastest growing plants on Earth. By comparing with other cores and using microfossils for dating, the period of *Azolla* growth was estimated to last 1.2 million years, from 49.3 to 48.1 Myr, a period known to geologists as early to mid Eocene. During this period the Earth, and also the Arctic was still hot. Estimates of the temperature of the seas using novel biomarker techniques (lipids in the cell walls of microbes that appear to show a remarkable temperature sensitivity that can be used to reconstruct sea temperatures) on the cores suggest temperatures around 10 °C or even higher. This would have pro-vided *Azolla* with a comfortable growing environment, including an increased hydrological cycle, where precipitation would have exceeded evaporation, and thus provided the background for the freshwater layers required. Based on analysis of the carbon con-tent of the sediments, and taking into account feedbacks with the carbon cycle in the oceans, scientists estimated a reduction of 55–470 ppm could have taken place through the appearance of *Azolla* in the Arctic Ocean in the Eocene. Of this the growth of *Azolla* itself contributed at least 40% to this carbon drawdown via net carbon fixation and subsequent sequestration.[10] According to the Dutch scientist Henk Brinkhuis, one of the investigators on the expedition, the relatively closed off Arctic Ocean at the time could have looked like today's Black Sea with organic material produced by *Azolla* sinking directly to the oxygen-deficient bot-tom where it cannot be decomposed, thus increasing the storage

[10] Speelman, E., et al., 2009

of carbon. This could be one of nature's mechanisms to counter high temperatures and CO_2 concentrations.[11]

Let us continue our time travel through geological history while Earth moves ever closer towards the icehouse. The transition around 34 Myr is known as the Eocene–Oligocene transition; it marks the final decline of a hothouse planet into an icehouse planet. Relatively small changes in the balance between the sources (volcanic eruptions) and sinks (weathering) had brought down the atmospheric concentrations to around 400–900 ppm,[12] an almost halving of the concentrations in the Eocene. This reduction in CO_2 provided the background for a large cooling of the planet, by 4–7 °C in the tropics and even more at high latitudes. Within about 0.5 Myr the Antarctic ice sheet had established itself, likely aided by changes in ocean circulation caused by the opening of the Drake passage between South America and the Antarctic. This was preceded by a long-term reduction in CO_2 that took about 15 million years. Once the ice sheet was established, it continued to grow, aided by feedbacks such as the increased albedo from ice, that reflects about 80% of the Sun's rays back to the atmosphere, compared to 5–20% from open water. This relatively quick establishment following a slow and gradual change is often referred to as a tipping point. The Earth in that case has entered into a different state (or equilibrium), from which it is hard to escape back to the original state.

We know relatively little about the next phase, the Oligocene climate. We know that the ice sheets in the Southern Hemisphere waxed and waned, probably in relation to the movement of the Earth around the Sun (more about this in the next chapter) and

[11] Eek, A., 2006.
[12] Lear, C., et al., 2020

that the CO_2 concentration in the atmosphere continued to drop further. Changes in the ocean gateways may have influenced the climate, but in general the term 'doubthouse', a stage in between the hothouse of the Eocene and the icehouse of the Miocene might be more appropriate. During this doubthouse CO_2 concentrations probably approached the modern values of around 400 ppm.[13]

Northern Hemisphere ice sheets appeared first during the end of the Pliocene and the start of the Pleistocene, around 2.58 Myr, when our first ancestors were probably roaming around the savannas of Africa. In this case the closing of the Panama gateway and associated changes in circulation, caused by tectonic movement, played a strong role. This period started at low CO_2 concentrations at around 300 ppm and provides the starting point for our next chapter on the evolution of CO_2.

[13] O'Brien, C., et al., 2020

CHAPTER 7

CARBON DIOXIDE AND THE WAXING AND WANING OF ICE SHEETS

W e explore the fantastic but elusive rhythm of the ice ages, their temperature and their CO_2 concentration during the last million years. The first Northern Hemisphere ice sheets had started to appear on Earth about 2.6 Myr. It was the founder of geology James Hutton, who first established the connection between gigantic boulders found at some places near the Alps and their possible transport by glaciers. But Louis Agassiz really hammered down the point in 1837 at a meeting in Switzerland where he postulated that large parts of the Northern Hemisphere could have been completely covered by ice sheets. Geologists soon started to produce maps of glacial extent, among them the US scientist Chamberlain, who mapped the extent of the Laurentide ice sheet in the Northern US. The geologists also started to realize the implications of the fact that ice consists of water. This water had to come from somewhere, and soon they estimated that a 100 m drop in sea level could have occurred. They also discovered that there had not been just one, but many ice ages.

To understand what was going on we need to go from the big boulders on the land to the sediments of the ocean where steady accumulation of material from above produces a rather precise historical archive of events. Drilling cores in this archive gives us a unique picture of past climate change. First, scientists analysed the species composition of the sediments for Foraminifera species (a group of unicellular organisms with an external skeleton of calcium carbonate) and then they looked at the isotopic composition of the shells. Both analyses found a regular change between glacial and interglacial periods. However, the key to understanding Pleistocene palaeoclimate was found by analysing foraminifera that did not live in the surface waters, but those that lived deep down, where temperatures were stable and reflected only the change in composition of the water and not of the temperature. Nick Shackleton and Neil Opdyke identified, by analysing the isotope composition of these so-called benthic foraminifera, 22 different stages, corresponding to 11 glacial cycles each with a glacial and interglacial.

But the next question was of course, what caused this rhythmic behaviour? For that we look towards the skies, or towards the astronomical theory of ice age forcing. The Serb Milutin Milankovitch must have loved doing sums as around 1920 he calculated how small changes in the path of the Earth around the Sun could have produced variations in radiation. To be more precise, on the suggestion of Wolfgang Köppen, the father-in-law of the discoverer of continental drift Alfred Wegener, he calculated these variations at 55, 60, and 65 degrees North for the summer, because in cold summers if snow does not melt, ice can form. He found that these low radiation, cold periods corresponded exactly to those of ice ages. It would however take until 1976 before Nick Shackleton could show that the three distinct peaks at 43,000, at 24,000, and at 19,000 years that follow from Milankovitch's calculations also appeared in the ocean sediment data. There was a connection between the skies and the deep ocean.

Arrhenius had used the CO_2 concentrations in the atmosphere to suggest that this could trigger an ice age. Drilling into the ice of the ice sheets provided a clue to the variation of CO_2 because the atmosphere gets locked into the ice and provides a near continuous record of past changes, much as the foraminifera provide a record of past changes in the ocean. The Danish scientist Willie Dansgaard and the Swiss Hans Oeschger analysed the composition of these bubbles in the ice and discovered changes in temperature and CO_2 that matched the geological periods of glacial and interglacials. Later drilling allowed reconstruction of up to 800,000 years, comprising several cycles. The analysis showed that warm interglacials usually had CO_2 concentrations of around 280 ppm and cold glacial periods of 180 ppm. These variations look like a plot of the heartbeat of the planet. What exactly causes this rhythm to occur is still something of a mystery, but complex interactions between ocean circulation, carbon cycle, and the Milankovitch cycles are likely to provide an answer.

Some twenty thousand years ago, Earth experienced one of its lowest values of CO_2 in the atmosphere since its origin. Yes, it had been going down during the icehouse conditions of the Pliocene, but touching 180 ppm was something new. Reaching this low value was part of Earth's rhythmic dance of waxing and waning ice sheets during the Pleistocene, 'the most recent period', named by Charles Lyell from the Greek words πλεῖστος (most) and καινός (recent). High values of CO_2 occur during the Pleistocene warm periods when the ice sheets retreated, and low values were the norm during cold periods in which the ice sheets expanded and moved towards the Equator to cover large parts of the forest and grasslands of the northern temperate zones. The discovery of this rhythm in CO_2 was possible through the study of ice core archives, buried deep below the surface of the ice sheets. The retrieval of the cores and their subsequent analysis presents us with one of the most fascinating stories of Earth science. What precisely influenced that dance and what factors determined the rhythm of the dance is still a matter of considerable debate.

Around 1800 the concept of an ice age was far from being universally accepted, let alone the role CO_2 played in its variability. James Hutton—who else but the founder of geology—had suggested some forty years earlier that the erratic and scratched boulders that were found in various places around the Alps were remnants of a past glaciation. Other than him, nobody really interpreted the isolated rocks in that way. That changed dramatically when the Swiss geologist Louis Agassiz took the floor on 24 July 1837 at the annual meeting of the Swiss Society of Natural Sciences in Neuchatel. Agassiz read his paper on the theory that the erratic boulders found in the Jura and Alps were not of local origin but had arrived with the movement of glaciers.[1] Based on work with his colleagues and mentors de Charpentier and Venetz, Agassiz set out the theory that the whole Northern Hemisphere, as far south as the Mediterranean and Caspian Seas, had been covered with a vast sheet of moving glacial ice. He also, and wrongly, maintained that the glacial drift found around Neuchatel came from the north and not from the Alps.[2] In fact, the glaciers of the Alps formed their own centre of glaciation in the cold periods. He coined the term ice age (Eiszeit) for these periods of

---

[1] Imbrie, J. & Palmer-Imbrie, K., 1979
[2] Wright, G., 1898

extreme cold.[3] De Charpentier and Venetz had also thought that the boulders were transported by glaciers, but Agassiz's fundamental insight was that it would have been so cold in the whole Northern Hemisphere that the ice would have covered everything up to its southern boundaries.

Given his youthful enthusiasm and belief in the correctness of his theory, his talk caused an uproar. During a fieldtrip to the heart of the Jura, the next day, it appeared, however, that he might have misjudged the support of his peers for his theory. A casual remark from Leopold von Buch about *'the stupid remarks made by some amateurs who had joined the group'*[4] reminded him painfully that the elite of the geological establishment still thought nothing of his ideas. The general theory was that the large boulders would have been brought when the last great (biblical) flooding had occurred. Charles Lyell offered a similar explanation, albeit with a little more ice in it, suggesting that they had been transported by icebergs. Agassiz's theory was heresy, if not against God, then certainly against the dominant geological theory of catastrophism. He would, however, not give up and spent the next few years collecting further evidence for his theory. Support came initially from the Oxford geologist and priest the Reverend William Buckland, whom he managed to convince of his theory during a field trip to Scotland. Buckland subsequently managed to quickly convince Charles Lyell, and after that things became somewhat easier for Agassiz. His *'Études sur les glaciers'* (1840; Figure 7.1) contained the new work he had undertaken after his 1837 lecture

[3] In fact it was his colleague Karl Friedrich Schimper who first used the term in a note published in the same volume of the *Actes de Sociéte Helvétique de sciences naturelles* wherein Agassiz's lecture was published. See Woodward, J., 2014.

[4] Imbrie, J. and Palmer-Imbrie, K., 1979

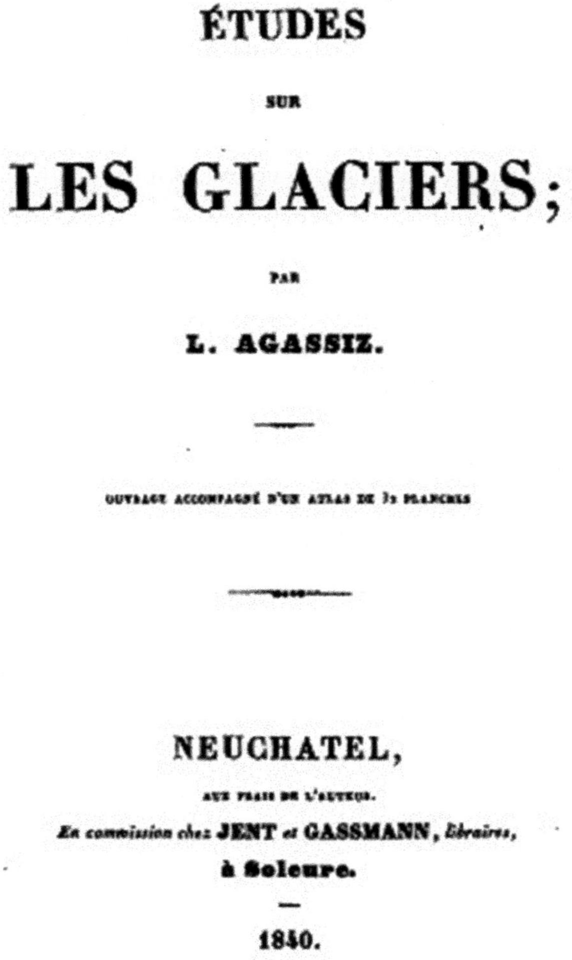

ÉTUDES

SUR

LES GLACIERS;

PAR

L. AGASSIZ.

———

OUVRAGE ACCOMPAGNÉ D'UN ATLAS DE 32 PLANCHES

————

NEUCHATEL,

AUX FRAIS DE L'AUTEUR.

En commission chez JENT et GASSMANN, Libraires,

à Soleure.

—

1840.

Figure 7.1 Cover of Louis Agassiz's study on glaciers from 1840.
Agassiz, L., 1840

and expounded the theory that the ice sheets had covered large expanses of the Northern Hemisphere. He dedicated the work to his mentors de Charpentier and Venetz. In 1846, in severe financial trouble because of failed business ventures, Agassiz visited the United States on a lecture tour which turned out to be a huge and popular success, although later tainted with accusations of severe racism. However, his expertise, particularly in fossils, was widely recognized and celebrated. In 1848 he accepted a professorship at Harvard. He started organizing and acquiring funding for a new museum of natural history. In 1860 his dream came true with the opening of the Museum of Comparative Zoology.

By the end of the century (1875) geologists were able to produce a sort of global map of glacier extent. This was based on the geologists' ability to recognize the remains of glaciers in various places from stratified deposits laid down by flowing water and interpret the irregular 'big boulder' deposits by the retreating ice that had set Agassiz on course for his theory. Working in America, T.C. Chamberlain produced a map of the Laurentide ice sheet that covered almost half of the current US. Soon geologists also realized that all that ice (i.e., frozen water) would have to have come from somewhere. They realized that there was in fact only one place where it could come from: the oceans. Hence the first calculations of the effect of ice sheets on sea level started to appear. They estimated a potentially massive 100-metre drop in sea level that would have drastically changed the shape and size of coastal zones. Importantly, by looking at repetitive layers, they also realized that there was probably not just one ice age, but a repetition of glacials (ice ages) interspaced by warmer intervals, the interglacials. This framework, in dire need of a theory to explain what caused the ice sheets to grow in the first place and the variations between glacials and interglacials, was of course what attracted Arrhenius and made him embark on his 1896 study (Chapter 2). Arrhenius' main interest was to see if small variations in CO_2 in the atmosphere could cause these shifts. We now know that he did indeed show that such variations could impact global temperatures by a sufficient amount to create glacial conditions.

Before we dive into the CO_2 variability during the Pleistocene, let us complete the story of the ice ages, by looking at the overall variability and timing. It was the Scottish geologist Archibald Geikie who argued in 1863 that deposits of glacial till that contained layers of peat in between showed that there had not been one big glacial period but rather several ones, interspaced by warmer intervals, the interglacials. The peat deposits between the glacial deposits could only have formed during warmer intervals between the glacial periods. In the early twentieth century, the main stages were named by geologists working in the Alps such as Penck and Brückner as Günz, Mindel, Riss, and Würm after the river valleys where the main evidence was located. However, geologists working in other environments named them usually after the rivers and localities where they were working. So, while the climate was getting alternatively colder and warmer and leaving global fingerprints of those periods, the geologists had not agreed on a uniform naming protocol. Take for example, the last interglacial before the current, it is known in the Netherlands as the Eemian, in the Alps it is known as the Riss–Würm interglacial, and in the United Kingdom as Ipswich. Moreover, it was very hard to put an exact date to the moraines found in the landscape, so no chronology of the different ice ages could be established.

At this point it is helpful if we leave the land for a while and move away from the big boulders and other ice-age induced land scars and look instead to the ocean for further clues. Scientists had noted, when taking simple samples of the ocean floor, that these were covered by blankets of sediment. The ocean environment is much more stable than the land: year after year these sediments accumulate on the ocean floor providing a precise record of climate. When scientists started analysing these sediments, they discovered that they consisted of tiny, mineralized relicts of plankton. Most of the calcareous shells on the ocean floor are the remains of planktonic foraminifera, which dominate the sediment in tropical to subpolar regions. In the Arctic and Antarctic waters, the relicts of siliceous organisms prevail—radiolaria, and diatoms. Knowing that some of these remains would have lived in cold and others in warm waters might provide a clue to determining the history of climate and ultimately the ice ages. One species of the forams, as they are colloquially called, is *Globorotalia menardii,* which was exclusively found in the layers of low-latitude warm waters. At high latitudes the species was notably absent. Using this species and radiocarbon dating it was possible to exactly date the transition of the last ice age to the present Holocene at 11,000 years.[5] The four authors—Ericson, Broecker, Kulp, and Wollin—famously stated in their last sentence of the conclusions: '*Further correlation of events both in the ocean and on land during this interval may lead to an understanding of some of the factors causing glaciation.*' They were right, but it was not as straightforward as they perhaps thought.

While Ericson continued analysing the cores, a monumental task given the roughly 200 cores that were taken each year by the Lamont research vessel *Vema,* an Italian postdoc Cesare Emiliani had meanwhile started analysing the isotopic composition of the foram shells in Chicago. This work stemmed from a suggestion by

---

[5] Ericson, D., et al., 1956

Harold Urey (see Chapter 4) that oxygen isotopes might give an indication of the past temperature of the ocean. Urey had found that uptake of water from the surroundings depends on the temperature of the water, with skeletons in colder waters containing more of the heavy oxygen isotope (^{18}O). That is, if for the time being we forget that the isotopic composition of water also plays a role. In any case, Emiliani started analysing initially bottom-dwelling foraminifera from Pleistocene deposits and quickly, realizing the potential, turned to Lamont's sediment cores. In 1955 he published his first results,[6] showing that in the Caribbean core he analysed he could identify seven stages, comprising three full glacial–interglacial cycles over the last 300,000 years. He suggested from his oxygen isotope measurements that the temperature during the glacial stage would have been 6 °C lower than in an interglacial. While the curve produced by Emiliani matched the one produced by Ericson, some teething problems remained and led to considerable debate between the two teams. One of the key questions was whether the variation in foram composition that Ericson analysed really did reflect only changes in the temperature of the water, or if it reflected some other causes such as, for instance, salinity or the amount of food present in the water. The isotopic composition of ocean water could also change during an ice age, complicating Emiliani's analysis (this was the issue that we blissfully ignored previously in the suggestion of Urey). In fact, it turned out that the variations that Emiliani observed were reflecting changes in the volumes of the ice sheets rather than ocean temperature because ice sheets contain frozen water that originates from snow. This is a somewhat complicated story that goes

---

[6] Emiliani, C., 1955

as follows. Precipitation near the poles is lighter in oxygen isotope content because the heavy isotopes have rained out during the transport of the water vapour from the tropics to the poles. Since it is easier for a light isotope to evaporate the remaining ocean water then contains more of the heavy isotopes. In a warm period such as an interglacial, the snow falling on the ice sheet is also bit heavier than during a cold period where more of the heavier isotopes have already rained out before the airmass reaches the poles.

The definitive answer to the question came from an analysis not of planktonic foraminifera, but rather from deep-dwelling organisms, the so-called benthic foraminifera. Forams in this deep and dark part of the ocean live in an environment where the temperature does not change much, is close to zero, and, importantly, where the temperature is hardly affected by the glacial–interglacial cycle. The shells of the forams thus reflect changes in the water composition rather than changes in the temperature. Part of the uncertainty surrounding the analysis of the deep cores resided in establishing a reliable chronology (what palaeoclimatologists call an age model). In their much-celebrated paper Nick Shackleton and Neil Opdyke[7] used the last time the Earth's magnetic field flipped (780,000 years ago) as one such critical age marker. From there they counted upwards and established a record of the previous years with the upper dates provided by radiocarbon analysis. The core they worked on was known as V28-238 and taken by the *Vema* vessel from Lamont. Imbrie and Palmer-Imbrie[8] call the resulting analysis the Rosetta Stone of Pleistocene climate. Importantly, they showed that the variation

---

[7] Shackleton, N. & Opdyke, N., 1973
[8] Imbrie, J. and Palmer-Imbrie, K., 1979

in oxygen isotope content between planktonic and benthic forms was very similar: proof that they responded more to the isotopic content of the surrounding water than to temperature. Figure 7.2 plots their final result. The different stages of warming, counting backwards from the top, are now known as marine isotope stages.[9] In total they counted 22 such stages, corresponding to 11 cycles, with a glacial and interglacial. The need to name ice ages after the location where they found was gone. A global record, numbering marine isotope stages from the top (our Holocene) down replaced the old geographical names.

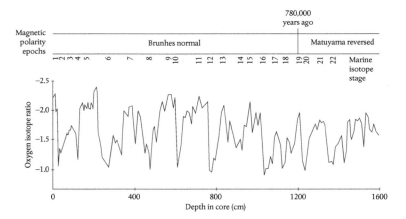

Figure 7.2 Analysis of the oxygen isotope in core V28-238 showing the variability of the late Pleistocene. At 780,000 kyr a magnetic reversal took place which helped to absolutely date the core; radio carbon dating was used for the more recent stages.
Redrawn from Shackleton, N. and Opdyke, N., 1973

[9] Emiliani's work had provided the basis for this scheme, but also for the somewhat anomalous counting of stage 3 and 4 which do not really represent a glacial–interglacial cycle but the numbering was kept intact by palaeoclimatologists (personal communication with Ganssen, G., 2021).

Having established the prime variation in temperature and the rhythm of the ice ages, the question now, of course, arises as to what is the cause of this variability? For that we need to look up to the skies, rather than down into the muddy sediments. We look at the work of a Frenchman, Joseph Adhémar, a Scotsman, James Croll, but above all a Serbian mathematician, Milutin Milankovitch. These three scientists developed what is known as the astronomical theory of climate forcing. Key to this is the fact that Earth does not exactly follow the same path around the Sun, but that this path and thus the exposure to the Sun has small variations. These variations can lead to less solar radiation being received at northern latitudes, for instance if the high latitudes are farther away from the Sun than in other conditions. If the solar radiation input were to be so low that snow having fallen during the winter would not melt in summer, an ice sheet could start to build up. Figure 7.3 illustrates three variations in the Earth's path. The eccentricity cycle describes variation that exists because the shape of Earth's ellipse around the Sun varies a little; another has to do with the alignment of the Earth's North–South axis to the plane of the Earth's orbit. This is called the obliquity cycle. The last important variation is the wobble of the axis, known as precession, which determines which part of the Earth is closest to the Sun. Calculating the final variability in radiation from these three combined effects without the help of modern-day computers is something hard to imagine. Luckily Milankovitch was able to draw on some previous work by a German mathematician Ludwig Pilgrim, who had done these calculations in 1904. After the first Balkan war in 1912, in which Milankovitch was drafted to deal with foreign correspondence, he decided to write down his initial calculations in three short books. In 1914 the First World

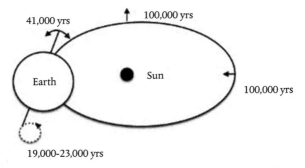

Figure 7.3 Earth's three cycles of eccentricity, obliquity, and precession. The years mentioned indicate the length of the average cycle.
After Dolman, H., 2019

War broke out, and again interrupted his work; this time he was taken as a prisoner of war, but subsequently he was released and able to spend the next four years calculating how much radiation the Earth would receive given his calculations of Earth's position and stability. These calculations were published in 1920 under the title '*Mathematical theory of heat phenomena produced by solar radiation*'. Celebrating 100 years of what is now known as the Milankovitch theory, the authors of a recent short memorial piece wrote that his work and his life in particular '*reveal a remarkable resilience in the course of developing one of the most beautiful theories in the Earth sciences*'.[10]

To turn his calculations into a proper theory, Milankovitch needed to know the real cause behind the triggering of an ice age: was it the low winter temperatures or average annual temperatures or just summer temperatures? And what latitude was critical? Milankovitch was helped here by Wolfgang Köppen, a

[10] Cvijanovic, I., Lukovic, J., and Begg, J., 2020

German climatologist who was interested in the distribution of vegetation over the planet. He quickly saw the relevance of Milankovitch's work when it was published in 1920. Köppen also happened to be the father-in-law of Alfred Wegener, whom we encountered earlier in Chapter 4 as the originator of the continental drift theory. Köppen suggested that it was the decrease in summer heat in the Northern Hemisphere that would be critical, as this would inhibit melting and thus initiate glacier growth. Milankovitch spent 100 days, day and night, calculating the radiation at latitudes 55 °, 60 °, and 65 °N for the last 650,000 years before he mailed Köppen the result shown in Figure 7.4. In it he was able to correlate the reduction in radiation with the occurrence of the main ice ages.

This work came in for severe criticism from geologists at the time of publication, despite Köppen's and Wegener's strong support in their 1924 book 'Climate of the Geological Past'. However, it acquired renewed attention when Shackleton's graph showed a similar cyclical behaviour. Shackleton teamed up with James Hays and John Imbrie to perform a spectral analysis on the foram and radiolaria data and the oxygen isotope records of the ocean sediments to see if they could also find evidence of cyclicity. A spectral decomposition, as it is formally known, shows whether there are preferred frequencies in the data record that can be interpreted as evidence of cyclicity. Published in 1976,[11] they identified three distinct peaks: at 43,000, at 24,000, and at 19,000 years. These corresponded with the main cycles of obliquity (43,000) and the two others at 24,000 and 19,000 years that could be identified as the precession cycle, with a major and minor variant. They also

[11] Hays, J., Imbrie, J., and Shackleton, N., 1976

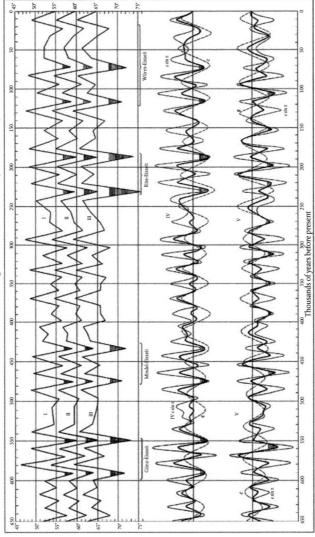

Figure 7.4 Milankovitch's calculations of radiation at 55 °, 60 °, and 65 °N. The curves show how radiation diverges over time—at 240,000 years ago the radiation received at 55 °, 60 °, and 65 ° N would be equivalent to that received at 76 °N today. The two lower panels show the contributions of the three cycles. The German names of Northern Hemisphere glaciations are also given on the x-axis.

From Köppen, W. and Wegener, A., 1924

found in the three indicators evidence of a 100 kyr cycle, but the response of the climate system to this forcing was non-linear, in contrast to the two smaller cycles. The amplitude of the forcing of the eccentricity is also smaller than that of the obliquity and precession cycles, so one would not expect a priori such a strong signal. The radiative effects of the obliquity and precession cycles work almost directly on the climate and thus show a linear response. In the case of the 100 kyr cycle this direct (linear) response was not evident in their calculations and they made the point that, rather than astronomical forcing, other factors could play a role here. This problem gains in significance as we go back in time more than one million years and realize that the progression of glacials and interglacials occurs at 40 kyr time intervals. This change, known as the mid-Pleistocene transition is still cause for debate among climate scientists and is known as the '100 kyr problem'. What is the forcing mechanism for this remarkable shift in phase from 40 kyr to 100 kyr? It is here that the astronomical theory provides no answers. One of those answers could be the behaviour of the ice sheets themselves, or, more precisely, its impact on the surface over which it lies. When ice sheets grow, they become so heavy that they literally push down the Earth; when they melt the Earth slowly recovers. We can see this process currently happening in Scandinavia where the average glacial rebound (as it is called) is of the order of a few mm per year, and in northwest Canada where it reaches almost 20 mm per year. Lowering the surface of the ice sheet exposes it to higher temperatures so that melt can increase. Impose then an increase in summer radiation and the effect may be enhanced. Computer models that consider all these processes predict the build-up of an ice sheet taking around 80–90 kyr with a much faster collapse taking only 10 kyrs.

Having gained some understanding of the rhythm of the waxing and waning of the ice sheets, the big question now is, of course, is it variation in CO_2 that plays a role in these ice age variations as Arrhenius supposed? To be able to answer that question, or at least to try, it would be nice to have a record of past changes in CO_2, preferably a record that can also be very accurately dated. We must thank the organizers of the International Geophysical Year for their foresight. They made it possible for a group of US scientists to travel to Camp Century, high on the Greenland ice sheet.[12] After five years, with much hardship, broken drills, redesigned equipment, and above all cold working conditions they managed, under the leadership of Henri Bader of the US Army's Corps of Engineers CCRL (Cold Research and Engineering Laboratory), to complete a core of about 1.4 km depth that covered the last 100,000 years. They also drilled ice cores at Byrd Station in Antarctica, where they retrieved a core of 2164 m in 1968. Subsequent analysis of the cores would provide our first glimpse of the atmosphere during the ice ages.

When snow accumulates, say in the middle of the Greenland ice sheet, the annual melt is less than the amount of snow that has fallen. If snow falls the next year, it will overlay the previous year's remains of snow and so, over a long time, we get an accumulation of layers. The weight of the snow will compact these layers; the deeper layers will be less thick than the top layers. In this process air will be pushed out, and ice will form, but some bubbles will remain in the layers. These bubbles contain a fingerprint

---

[12] Langway, C., 2008

of the atmosphere when they formed. In other words, if we are able to determine the age of the layers and analyse the bubbles we can get a rather precise reconstruction of the amount of CO_2 that was present in the atmosphere when the snow fell. Unlike other climate proxies, we have a direct observation of the past by analysing the air of the ice cores. This in short, is the beauty of ice core research.[13]

The real story, as often in science, is much more complex. Snow not only gets compressed, but the ice also moves laterally, and at different speeds. The pressure of the ice makes it move sideways, towards the coasts of Greenland, where glaciers dispose of the ice into the surrounding seas. At the top of the ice sheet this movement away from the top slightly lowers the surface, just enough to fill it up again with new snow—that is, provided the climate does not change that much. All one now needs is an indication of the age of the layers. Here, at least for Greenland, the seasonality of the input from the Sun comes to our assistance. During the winter the Sun is absent over Greenland and it is dark almost 24 hours a day. The opposite occurs during the summer. When the Sun shines, the surface snow cover sublimes; moisture then forms a frosty layer on the snow during the short night when temperatures drop again. This process makes the summer layers distinctly less dense than the cleaner winter layers. We thus find a clear alteration of darker and lighter layers that is indicative of a full year. The layers remain visible in the snow, the compacted version called firn, and ultimately in the ice. On Greenland, it takes about 200 years to form ice, which by then is about 60 metres down in the ice sheet. The cores also contain any large disruptions of ash from volcanic

---

[13] For a wonderful introduction into ice core research read Alley, R., 2000.

eruptions that may help to date a particular layer. Beyond 50,000 years old this visual dating becomes rather problematic and other techniques such as water isotopes are needed.

For this we need to turn to the Danish geophysicist Willie Dansgaard (1922–2011), who in 1952 had the brilliant idea to use an empty beer bottle to collect rainwater to investigate the isotopic composition of rain and determine whether it changed with temperature of formation. He was very lucky that during that particular weekend in June 1952 a very clear frontal situation passed over Denmark: first cold rain before the arrival of a warm front, then cold rain at the front, and then rain that formed at decreasing altitudes during the passage, and thus from relatively cold high up to warmer at lower altitudes. When a subsequent cold front arrived, he had new observations of precipitation from high-altitude colder air. When he determined the isotopic composition, he found clear evidence of changes in the occurrence of the heavy ^{18}O isotope. But how to turn this insight into a reliable climatological thermometer?

While the International Atomic Energy Agency and later WMO saw great relevance in this idea and started to work with Dansgaard to setup a global network to collect rainfall for isotopic analysis, Dansgaard's attention was now also drawn towards Greenland. He realized that his method might also provide insight into the temperature of the air when ice was formed. The reasoning was clear: *The most important aspect was implied in an "internal reasoning" that the present* δ *(the deviation in isotope ratio from the* sample compared to a fixed standard) *temperature relationship for cold regions might also be valid when going back in time. In other words,* δ *in old water might reflect the climate at the time of formation of the water. Now, where do you find old water? In glacier ice. And where do you find old glacier ice? In Greenland.*"[14] Dansgaard teamed up with the zoophysiology expert Per Scholander, who had published a note in *Science* in 1954 suggesting that diffusion of gases (CO_2) in glacier ice was so low that the bubbles remaining would represent atmospheric conditions at the time of formation. So, Willie Dansgaard and Per Scholander went to Greenland to sample ice from icebergs on a ship called the *Rundøy*. The events during this 'bubble expedition' are delightfully retold by Dansgaard in his memoirs and include dangerous accidents, socializing with local Greenlanders, and hard work on the ship to collect the ice from icebergs. They were able to show that ice from older icebergs, that came from colder areas further inland from the ice sheet coast, had a lower amount of $\delta^{18}O$ than ice that was younger. However, there was one glitch—melting would release gases from the bubbles, and, indeed, the bubbles Scholander and Dansgaard were looking for had been contaminated by probable summer melt, making the composition analysis unreliable. If one wanted to find old atmospheric air one had to turn to glaciers. Dansgaard was able to get his hands on cores from firn that were collected by the Expédition Glaciologique International au Groenlande (EGIG) that gave him access to several 10–20 m deep cores. This produced the first isotope climate record based on firn. It ran from 1910 until the mid-1950s and it showed a warming from 1920–1945 and a subsequent cooling in the two deepest cores, with the cooling trends being similar in all five cores taken.

It was now time to take on a more serious challenge. The US Army Cold regions Research Engineering Laboratories (CRREL) had started to drill a large deep core at their site at Camp Century,

---

[14] Dansgaard, W., 2005

Figure 7.5 Thermal coring rig in Trench 12 at Camp Century, 1964. The two men are standing next to the ice core increments in carriers. The composite building at the rear is the drillers' workshop.
From Langway, C., 2008

about 20 km from Thule (Figure 7.5). The camp was a massive undertaking, started during the cold war and for some time even housed a nuclear reactor for its energy supply. It ran all year around and during summer it could hold up to 250 people housed in 32 buildings dug into the firn. It was an environment worthy of a James Bond movie. During the period 1960–1966 and under the supervision of Lyle B. Hansen, they drilled several cores of several hundred metres. The last core reached the bedrock below the ice sheet at 1388 m. Dansgaard put in a proposal to measure isotopes along the full length of the core. The overall responsible scientist from CRREL, Chester Langway, agreed. Dansgaard now had access to a core that probably covered a full glacial cycle. To determine the age (radiocarbon cannot be reliably used for dating beyond 40,000 years) required modelling the thinning of the layers with depth that we mentioned above, which arose because of the glaciers moving outwards from the centre. Sampling every two metres they were surprised to see that they found a peak in temperature at around 100 kyrs—might this be the previous interglacial, the Eem? The results confirmed the earlier analysis on the smaller firn cores with a climate optimum at the beginning of the twentieth century and subsequent cooling. But they also found evidence of the little ice age and the medieval warm period. The last glacial clearly stands out as well, from about 70–10 kyr. The Rosetta Stone of past climate variation was found! But how about CO_2?

As with Dansgaard, the core also attracted the attention of the Swiss scientist Hans Oeschger, who in 1966 and 1967 published the results of radiocarbon dating the cores. The collaboration between the Swiss, Danes, and CCERL led to the development of the Greenland Ice Coring Project (GISP). The Russians had also started ice coring in 1955 and would eventually produce the most famous of ice cores, the Vostok ice core in Antarctica that would reach a depth of 3769 metres where it hit the interface with Lake Vostok and drilling was stopped. This core that would dramatically change our understanding of the interplay between CO_2 and climate. A European consortium (EPICA) started drilling at the so-called Dome C in the east of Antarctica in 1997, where they reached bedrock at 3260 metres, providing a record of slightly more than 800,000 years.[15] The 800 kyr limit of the ice is something of a disappointment: knowing from the marine record that around that time the rhythm of the ice age changed from 40 kyr to 100 kyr, the hunt is now on for cores that extend into the 40 kyr era. The EPICA consortium started in 2021 to drill 'beyond EPICA's Oldest Ice'. One of the big questions, of course, is what drives the apparent synchronous rhythm of CO_2 and temperature in the Pleistocene, and how and why does that change in the mid Pleistocene?

Oeschger was initially interested in using the CO_2 trapped in the bubbles to date the core with radiocarbon. Initially, they used melting techniques to extract the CO_2 from the ice, described in a seminal paper published in 1980 analysing two cores, one from Camp Century and another from the Antarctic station Byrd, suggesting that *'For both cores, the values for the CO_2 concentration of the first fraction, considered to best represent the atmospheric composition, show lower values during glaciation than in the Holocene, with a minimum before the end of glaciation. Low CO_2 concentrations in the first fractions (200 ppm) of certain samples are a strong indication that the atmospheric CO_2 concentration during last glaciation was lower than during the postglacial. These low concentrations indicate that, at that time, CO_2 concentration in the atmosphere could have been lower than today by a factor of 1.5.'* At the same time, in Grenoble in France, Robert Delmas had started to realize that this technique might cause chemical reactions in the ice with an acidic constituent that would affect the CO_2 concentration in the measurement. They developed a dry extraction technique using ice crushers at -40 °C. Similar to the

---

[15] See Jouzel, J., 2013

Swiss they found that during the glacial period before our current Holocene, the CO_2 concentration in the bubbles was about two-thirds of the atmospheric value at the time of measurement: 200 ppm compared to 300 ppm.

This initial picture of glacials with low concentrations of CO_2 and interglacials or warm periods with high CO_2 was dramatically confirmed when the ice core from Vostok was fully analysed. Over the last 400,000 years not only had the ice rhythmically waxed and waned, but it was also (within the limits of how well we are able to determine the age of the ice and the bubbles) synchronized with the variation in CO_2 concentration. High concentrations up to 290 ppm occurred during warm periods and low values during cold periods of around 180 ppm. Figure 7.6 shows the last 800,000 years from the EPICA Dome C coring.

Figure 7.6 shows a similar forcing to that we saw earlier in the marine record at 23, 41, and 100 kyr, statistically similar to the Milankovitch forcing of Figure 7.4. But while it is possible to think of a direct link (correlation) between temperature and radiative input, it becomes much harder to envisage a causal relation between the Milankovitch forcing and CO_2. The big question now was, how does the variation in temperature and CO_2 interact with the Earth System and the Milankovitch forcing? The changes in radiative forcing are generally too small to generate strong growth of ice sheets, so, after an initial start, amplification mechanisms need to be invoked. The first of those is the enhanced reflection of radiation by the white ice sheets, which would lead to lower temperatures; the second could be the role of CO_2 in this amplification.

Apparent synchronous rhythm…? Not quite. Around 450 kyr the maximum CO_2 concentration does not get above 260 ppm,

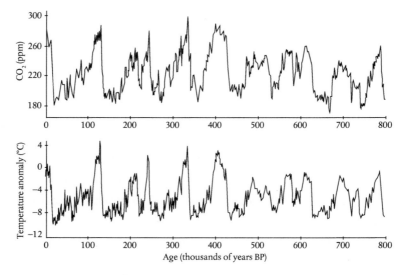

Figure 7.6 CO_2 results from Antarctic ice core Dome C with temperature anomaly (representing the difference between the actual temperature and the mean temperature of the past millennium) over the past 800,000 years.
Redrawn after Jouzel, J., 2013 and Lüthi, D., et al., 2008

whereas closer to us in time, the maximum is generally close to 280–290 ppm. Also, the last but one interglacial in the record, known as marine isotope stage 18, at 730 kyr reaches values only as high as 210 ppm. The corresponding temperature variations are similarly low. Shackleton, whom we encountered before as a sharp mind, analysed some of the ice core and sediment data together and found importantly that the ice-volume component (as obtained from the [18]O record) lags behind the changes in CO_2. Changes in the ice sheet volumes thus do not cause changes in atmospheric CO_2, but rather the other way around. Also, when looking at a finer timescale, it appears that in the Antarctic, CO_2 lags temperature by a few hundred years. In contrast, in the

Greenland cores, which do not contain so many ice age cycles as the Antarctic cores, temperature appears to follow the increase in CO_2. What arises is a far more complex picture than can be explained by a single causal factor, be it Milankovitch cycling, CO_2, or ice sheet decline. An adequate explanation of this multitude of causal factors is still missing; it could well be described as the holy grail of palaeoclimate ice age research. The currently most favoured theory involves changes in ocean circulation, the ocean carbon cycle, and nutrient availability in the Southern Ocean. It is likely that the ocean plays a key role, as this is the only reservoir that can absorb and emit over thousands of years enough carbon to change the atmospheric concentration. For instance, when ice sheets grow, they would cover large areas of forest and vegetation in the Northern Hemisphere, providing a reduction in the carbon sink strength. This reduction in uptake is estimated to be small, partly through compensating mechanisms in the ocean, where most of the carbon would be buried.[16]

During the late Pleistocene it is likely that the surface-to-deep exchange of ocean waters was somehow weakened in the Southern Ocean's Antarctic Zone. Under today's (non-glacial) conditions water that has sunk at the Arctic is transported deep in the ocean through the Atlantic Bottom Water current to the Antarctic where it wells up near the coast. Water near the South pole sinks by cooling and forms a flow of water towards the Equator that is called Antarctic Bottom Water. These flows are part of what is popularly known as the conveyor belt, an intricate system of ocean transport operating at the timescales of thousands of years. This system is thought to change during glacial

[16] Sigman, D. & Boyle, E., 2000

to interglacial transitions leading to a multitude of changes in the Southern Ocean. One of those changes is the leakage of carbon from upwelling ocean waters. This is the 'normal' situation during an interglacial, with the CO_2 becoming available during this leakage blown away towards the Equator. During a glacial period this situation changes and the leak is partly blocked. The reduced leakage of deep ocean carbon dioxide contributed to the lower atmospheric carbon dioxide levels of the ice ages. Other factors that could contribute to reducing the CO_2 output to the atmosphere are increased ice cover and a more efficient biological pump (algae taking up CO_2) that was able to consume most of the upwelling nutrients or changes in the alkalinity of the ocean.

The ice ages show a remarkable rhythm of a slow build-up of the ice sheets over 80–100 kyr, which then experience generally a relatively rapid demise in 10–20kyr. The timing of the onsets of the major ice ages correspond remarkably with minima in Northern Hemisphere summer radiation arising from the obliquity, precession, and eccentricity cycles calculated by Milutin Milankovitch. However, these changes are too small to fully explain the cycle, so we need to look for additional mechanisms. One of these could be the concentration of CO_2 in the atmosphere as observed in the bubbles of air in the ice cores. CO_2 during ice ages is about 100 ppm lower than that during interglacials. Explaining this variability invokes changes in deep ocean circulation and biological activity in the ocean. The full answer to these riddles, however, is still out there, maybe buried forever, deep below the surface of the Arctic and Antarctic ice sheets. In the meantime, Earth had reached what was probably its lowest CO_2 concentration in recent geological history. That changed dramatically when humans started to use fossil fuels, but that is the subject of the next chapter.

CHAPTER 8

HUMANS, FIRE, FOSSIL FUEL, AND THE RISE OF ANTHROPOGENIC CO$_2$

We discuss the appearance of humans on the scene, first as a factor changing the surface of the Earth through deforestation, largely by fire, and later by discovering the steam engine and the use of fossil fuel burning for power generation. At the beginning of the Holocene, 10,000 years ago 71% of the Earth's habitable land surface was covered by forests. Small-scale clearing by humans took place mostly for fuel wood and agriculture. Around 1500 large areas of forest were cleared in Europe, later in other parts of the world such as Latin America and Asia. Current estimates suggest that we now have cut or burnt down roughly 50% of the original forest areas, so we are left with only 38% forest cover remaining. The carbon of the trees ends up in the atmosphere when trees are burnt or left to decompose, in total about 200 Gton C, of which most was emitted after 1850 when deforestation was becoming easier by the availability of new technical means such as steam power. Historically, changes in atmospheric CO$_2$ were caused by the emissions from deforestation and land-use change; today these emissions still comprise about 10% of the fossil fuel emissions.

The rise of steam power after 1870 changed the historical picture of emissions as now increasingly coal was mined and burnt to generate power. By 1900, 70% of power generation was through steam power, with only 30% from humans and animals. Before 1950 most of the fossil fuel emissions came from coals, the share of oil and gas increasing after that. This is important as coal produces more CO_2 emissions per unit energy than oil and oil produces more than gas. Up to 1950 roughly 90% of the emissions came from Europe and the US. In the latter part of the twentieth century CO_2 emissions from China, the rest of Asia, and some countries in Latin America picked up. In 2006 China became the largest fossil fuel related CO_2 emitter, surpassing the US. The western world, i.e. the US and Europe, is however responsible for over 50% of the cumulative emissions since industrialization. The world's Gross Domestic Product, global primary energy consumption, and carbon intensity drive the global emissions, while each of these variables can differ between regions.

Of the current emissions roughly half end up in the atmosphere, causing the main rise in atmospheric CO_2 concentration. Both the ocean and the land currently take up the other part of the emitted CO_2: the land biosphere around 30%; the ocean a little bit less, around 25%. This free hoover service helps by taking up a considerable part of the CO_2 coming from fossil fuel burning, but it is unclear how long it will last before the land and oceans saturate and start turning from a sink into a source. So far, however, the service appears to be holding…

Human history began with fire. According to Greek mythology, the supreme God, Zeus, had used fire to great effect in the classic war of the young, aspiring Olympian Gods against their old predecessors, the Titans. In an act of pity for the feeble humans, Prometheus, fighting alongside the Titans, stole the fire and handed it to humankind. Prometheus was severely punished for his deed by Zeus, who had him tied to a mountain in the Caucasus where an eagle picked away at his liver during daytime. The liver that was ripped apart by the bird would grow back to its original shape during the nighttime, after which the torture would start again. Prometheus, evidently feeling rather sorry for himself, in the words of Aeschylus, said: *'For it is because I bestowed good gifts on mortals that this miserable yoke of constraint has been bound upon me. I hunted out and stored in fennel stalk the stolen source of fire that has proved a teacher to mortals in every art and a means to mighty ends. Such is the offence for which I pay the penalty, riveted in fetters beneath the open sky'*[1]. In the Promethean perspective, fire is a technological power, it aids humans to burn land, forge weapons from steel, and produce art. Indeed, over millions of years, fire has helped humans to cook, stay comfortably warm, develop art and technology and, of course, burn seemingly endless resources of fossil fuel. By harnessing the force of fire, human progress was unbound.

Historically, there is a tight relation between human development and energy supply. Initially, more energy supply aids development until at some point, it reaches a stable plateau where more energy does not necessarily contribute to more development of the things that matter most such as life expectancy, adult literacy, and per capita GDP[2]. In ancient civilizations, fire and human and animal labour were the prime energy sources, with windmills and water mills arriving late on the scene, at around 200 to 900 AD. But even by as late as 1800 it was estimated that in England 87% of all power was still generated by humans and animals.

[1] Aeschylus & Smyth, H., 1926
[2] Smil, V., 2017

Fire in the environment has been around much longer than humans. Fire is a key component of many natural ecosystems that experience natural cycles of burning and regrowth and fire impacts affect global biogeochemical cycles that are closely linked to climate[3]. Distinguishing between the human component and the natural is, however, very difficult. Remains of burning exist as fossilized charcoal and indicate that natural wildfires might have begun soon after the first appearance of terrestrial plants some 420 million years ago[4]. Next to vegetation to act as fuel, combustion also needs oxygen (Chapter 3). Fire can only occur when atmospheric O$_2$ concentrations are higher than 13%. In Earth's history, fire intensity has generally increased when oxygen levels increased. Many coals of the Permian period (roughly 300 million years ago) contain evidence of fire in the large quantities of charcoal. These remains were laid down during a period when atmospheric oxygen was thought to have exceeded 30%, compared to the 21% of recent times (note, locking away large amounts of carbon from circulation can free oxygen as we have seen in Chapter 3). After that period, oxygen levels have appeared more or less stable at around 21%.

Humans started to appear only recently in the geological record. While there is continued debate about what was the first human being and where he or she exactly appeared, it is relatively safe to say that roughly 2 million years ago humans, as bipedal animals, made their presence notable in the records in the savannas of Africa. The first evidence of human fire in the form of burnt sediments and stone tools is dated around 1.6 Myr. The timing of human development in its modern form as *Homo erectus* is

[3] Dietze, E., et al., 2018
[4] Bowman, D., et al., 2009

set at around 1.7 Myrs, remarkably close to the earliest fire finds. From around 400,000 years ago (in the middle of the Pleistocene period, see the previous chapter) more archeological evidence is available from other sites, in Europe, the Middle East, and parts of Asia, near to Africa. Initially, it is likely that humans, or their predecessors, made rather opportunistic use of naturally occurring fire in the savannahs of Africa. They foraged for food that could have become more visible after fire, or even started to eat cooked food. Managing a fire is, however, something different from opportunistically exploiting it. The former requires the ability to ignite fire, control its intensity and length and transfer to other places—a complete new set of skills. Around 120,000 years ago, evidence suggests that humans had mastered these important skills, so that they could heat stones and maintain and run fires in a domestic environment[5]. The long-term consequences of this mastering of fire were to be enormous. Fire, together with wood harvested for fuel and timber provided a stimulus for a more agricultural organization of societies[6].

Making another leap in time, towards a few thousand years ago, the use of fire in metallurgy started to appear. Humans were now capable of cupellation, separating precious metals such as gold and silver from their ores. Interestingly, one can find traces of lead in ice cores that testify to this. Later peaks indicate the use of coins, the exhaustion of Roman lead mines, and later the production of silver in Germany and in the New World by the Spanish. By now humans had mastered fire, they could contain it and use it

---

[5] Bowman, D., et al., 2009
[6] For a magnificent treatment of deforestation throughout human history, Williams, M., 2003. The book gives an extensive overview of human impact on the environment through deforestation through the ages.

in industrial processes, something the early hunter-gatherers and agriculturalists were unable to achieve.

At the beginning of the Holocene, 71% of the habitable land was covered by forest; this was reduced to 57% by ten thousand years ago. Five thousand years ago, at the mid-Holocene, that was reduced further to 55% (this may sound small, but it is a change of 0.2 billion hectares). Initially most of the deforestation was for fuel wood and clearing for agriculture. This would have serviced small settlements using slash and burn techniques, or for them to simply 'nibble' away at the edge of the forest. While the Greeks and Romans cleared some land and started to use trees for timber, the age of the great clearings, 'l'Âge des grandes défrichements', took place in Europe in the twelfth and thirteenth centuries. This was, at least for France and Germany, the age of land reclamation, in which religious orders and nobility played a critical role in converting forest land to agriculture. It is estimated that by the year 1500 almost half of the original forest in Europe had been cleared. In the words of Michael Williams: *'By any calculation, the medieval European experience must rank as one of the great deforestation episodes in the natural world'*[7]. This seemingly limitless deforestation and expansion of settlements was abruptly stopped when medieval Europe was hit by the bubonic plague, or Black Death. It wiped out one third, up to a half in some places, of the European population between 1346 and 1353, with a total death toll approaching 20 million people out of a total population of 74 million. Across wide areas of Europe settlements became abandoned and the forest started to regrow. There is a remarkable correlation between these large-scale pandemics and the concentration of CO$_2$ in the

---

[7] Williams, M., 2003

ice cores that suggest that the regrowing forests drew CO_2 from the atmosphere to reduce the Northern Hemisphere concentration by about 4–10 ppm[8]. This theory is controversial because a correlation does not necessarily imply a causal mechanism. In other parts of the world, particularly in China and Russia, deforestation also took place for land clearing and timber, but our ability to document those changes in comparable detail to those happening in Europe is pitifully small.

Converting forest land to cropland and grazing was aided by fire as an efficient clearing tool, but use of trees for timber started to play an increasing role in the deforestation of Europe. In particular, the next two centuries after the Black Death saw a further decline in forest area as timber was required to build towns, cities, cathedrals, and ships with which to explore the rest of the world. At the same time the world's trade became more globalized with the discovery of the Americas in 1492 and further exploration of the world in the Orient by the richer countries of Europe. In the seventeenth and eighteenth centuries large amounts of forest in the US were cleared by immigrants from Europe. Yet by 1700 only about 5% of the world's forest had been cleared. Between 1700 and 2018 (Figure 8.1), however, almost a tripling of that rate took place (14%). *The power and progress of industrial technology radically altered the pace and nature of the life and livelihood of a large number of people, and the character of their society. Europe and its appendages overseas — North America, Australia, New Zealand, and selected parts of Latin America — rapidly followed suit. In Asia, Africa, and the bulk of Latin America, aspects of the same technology, coupled with exploitation and domination by Europeans, had equally dramatic and ultimately devastating effects,*

[8] Ruddiman, W., 2005

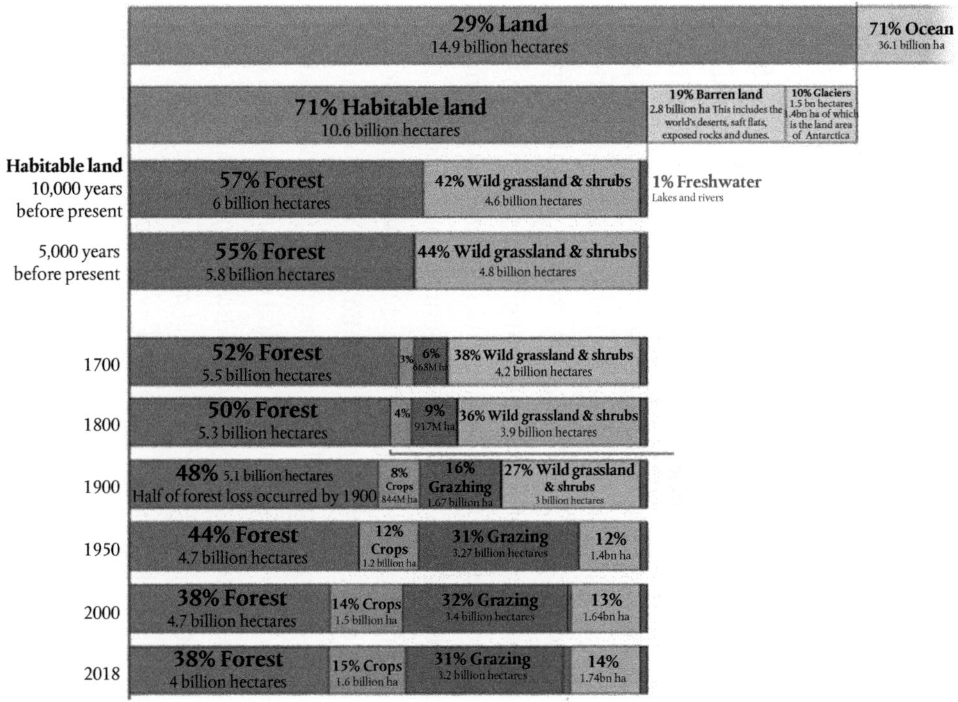

Figure 8.1 Change in land cover in the past 10,000 years.
From Our World in Data

overturning ancient ways of life, opening the way for a new global economy, and unbalancing world relations. Everywhere, land cover and land use experienced a dramatic transformation, of which, perhaps, the forest was affected most of all, followed in time by changes in grasslands. It was, as we have said before, the biggest change to the vegetation of the world since the Ice Age"[9]. Part of the latter 'progress' was undoubtedly associated with the availability of steam power to which we will come in an instant. At the beginning of the eighteenth century, forest clearing for agriculture (grazing and crops) took off more dramatically with the final result that we now have only 38%, roughly half, of the habitable land left as forest, with grazing and arable agriculture making up 36%. There is only 14% of wild grass and shrubland left, compared to 42% 10,000 years ago. In the process two billion hectares of forest, one third of the original cover, have been lost.

Trees contain carbon, so felling or burning them frees carbon. When timber is being used in building the release is slow as the structure slowly degenerates. When forest is burnt, the carbon is immediately released as CO_2 into the atmosphere. In principle, if one knows the carbon content, multiplying it by the area gives the amount of carbon lost. In practice there is a little bit more to this calculation. Carbon density tends to vary with tree species such as between coniferous and deciduous. It also sits in the soil. To make historical reconstructions one also needs to estimate the rate of settlement and population growth. All this information put together allows a reconstruction of the amount of CO_2 lost by land conversion, including grassland change to agricultural cropland. The German researcher Julia Pongratz first reconstructed these past emissions[10]. The first thing she notes is that the emissions of

---

[9] Williams, M., 2003
[10] Pongratz, J. & Caldeira, K., 2012; Pongratz, J., et al., 2008; and Reick, C., et al., 2010.

the period after the industrial revolution are almost twice those before 1850 (116 versus 62 Gton C). In Latin America emissions after 1850 are much larger than in the previous period (39 versus 5 Gton C). In Europe, as noted earlier, the emissions due to land-use change are much smaller after 1850. Before 1850 the bulk of the emissions came from Europe, South Asia, and China (57%); after 1850 Latin America, South East Asia, and Tropical Africa made up the bulk of the emissions (59%). Clearly deforestation and land-use change had shifted its geographical focus from the temperate zone to the tropics. The current deforestation rate is of the order of 1.2–1.5 Gton C per year, with most of that happening in the tropics.

We must now ask how much of these emissions stay in the atmosphere? It is here that we encounter in full the beauty of the carbon cycle on Earth. To investigate that question, one needs a model with an active biosphere and ocean that can indeed respond to increased carbon emissions. Well, it turns out that for the period before 1850, the biosphere responded actively to these changes and 48% of the emissions got taken up by the land. With the ocean taking up 31%, eventually, only 21% of the 62 Gton C of emissions remained in the atmosphere. This 13 Gton C is responsible for about 5–6 ppm increase in atmospheric concentrations. The main reason for this is that the slow, almost linear increase in emissions due to land-use change gives the biosphere and ocean time to accommodate. While this change in atmospheric concentration had negligible effects on the global temperature, the 5–6 ppm increase is higher than the natural variability within the Holocene. The impact of humans on the atmosphere became visible well ahead of the industrial revolution.

The picture changes dramatically when humans started to use fossil fuel for heating and importantly, to produce mechanical labour. In the eighteenth century, the steam engine appeared on the scene with its sheer endless capacity to harness the power of fire and make man independent of human and animal labour. The development of the steam engine was made possible by someone we have encountered earlier as the discoverer of 'fixed gas', CO_2, the Scotsman Joseph Black (Chapter 3). Black's love of experimentation led him to investigate the properties of water, ice, and water vapour, in particular how much heat was required to move liquid water to one of its other phases. He had noticed that melting ice remains at the same temperature and that a vessel of boiling water also does not suddenly transform all its contents into water vapour. It is the hidden heat, the latent heat of water as Joseph Black called it, that ensures that boiling water can absorb large amounts of heat without raising its temperature. If this was not the case *'the undeniable consequence of this would be an explosion of all the water with a violence equal to that of gunpowder'.*[11] Black had laid the foundation for the thermodynamics that would allow James Watt to further perfect the steam engine. Initial designs of the steam engine stem from the late seventeenth century where they were designed and used primarily to pump water, but it was the design of Thomas Savery early in in the eighteenth century that set things truly in motion[12]. After that the blacksmith Newcomen made the first useable steam engine that saw applications in pumping water out of mines. A redesign of Newcomen's model gave James Watt in 1779 the patent on which most steam engines were based afterwards: *'A new invented method of lessening the consumption of steam*

[11] Cited in Rosen, W., 2010
[12] Rosen, W., 2010

and fuel in fire engines'. Watt made several key innovations that improved the efficiency of the engine.

But overall, the uptake of steam engines outside the use of pumping water in mines was still small. The fact that they were primarily used in mines was also partly a result of the fact that they needed coals to produce the fire itself. The full impact of the steam engine came only in the second half of the nineteenth century, the reason why the uptake was relatively slow being a matter of considerable dispute among historians and economists.[13] In any case the standard depiction of the steam engine as designed by James Watt immediately causing the industrial revolution is incorrect. In other words, the number of chimneys spewing out CO$_2$ remained small in the early nineteenth century in England but picked up in the second part of the century. By 1870 one-third of the coal produced was used in steam engines, and the cotton factories of Lancashire alone were responsible for more burning of coal than the total of southern and eastern Europe put together. Part of the reason for this was that coal production had dramatically increased in Lancashire. In 1870 the emissions reached about 60 Mtons of carbon, almost double that of twenty years earlier. In fact, the growth curve of these early industrial emissions looks very similar in shape to the one that we have experienced over the last fifty years or so.

By 1900 steam had taken over and human and animal labour only amounted to 27% of total energy use. Coal production became increasingly efficient with the advance of the early steam engines that kept the mines from flooding[14]. Not all of continental Europe was able to follow England's conversion from organic (human and

[13] E.g. Malm, A., 2016
[14] Wrigley, E., 2013

animal power and burning wood) to an energy source based on fossil fuel burning. In the Netherlands peat burning provided an early equivalent of fossil fuel burning, contributing to the Dutch 'Golden Age'. Eventually, most of Europe and then later the rest of the world followed, but there remained a large proportion of countries with forests that still relied on wood burning (as numerous developing countries still do). In 1850, in Sweden, the Netherlands, Italy, and Spain, the use of muscle power was still 25, 38, 41, and 50% of the total energy use, respectively. Firewood made up 73% of total energy use in Sweden and 51 and 46% in Italy and Spain, respectively. In the Netherlands 10% was made up by wind and water energy, 41% from fossil fuel, mostly peat[15].

Fossil fuels are a way to store the radiative energy coming from the Sun. Peat and coal arise from the slow transition of plant material under pressure and heat, while hydrocarbons such as methane and oil are formed mostly from marine phytoplankton, zooplankton material, and algae. The formation of coals represents a concentration of carbon, from an initial 50% in plant material to 100% carbon in anthracite, while bituminous coals contain 85% carbon. In gas (mostly methane) the carbon content is about 75%. Bituminous coals make up the bulk of coal extraction globally and are notorious for the air pollution they produce when burnt[16]. The energy density, the amount of energy produced per kg fuel, is highest for anthracite and lowest for lignites (brown coal) with bituminous coal being intermediate. According to Vaclav Smil[16], the origin of coal extraction goes back to the Han dynasty in China (260 BCE–220 CE) where it was used in iron

[15] Warde, P. & Marra, A., 2007
[16] Smil, V., 2017

production. In England, Wales, and Scotland at places where coal was found near the surface it was used, and sometimes still is, but only near to the outcrops and, rather unexpectedly, by poor people, who could not afford to buy wood. The first records of coal extraction in Europe are from Belgium in 1113. First signs of trade in coal started around the twelfth century, with first exports from England to France in 1325. Declining wood harvest led in the fifteenth and sixteenth centuries to increasing use of coal in England: almost all the country's coalfields were opened up between 1540 and 1640. All the extraction was by human labour, sometimes aided by animal labour in the form of donkeys and horses. In the rest of Europe, coal was mined in Northern France, the Liege and Ruhr areas in Belgium and Germany, and in Bohemia and Silesia. In Russia the dominant energy source was still wood. Smil estimates that almost 75% of the total energy used in Russia in 1913 was still from wood burning. China, while using coal as one of the first in the melt process of metal, was relatively slow to convert its main energy source from wood to coal. Only in 1965 did coal use overtake the burning of wood. All in all, large differences in coal use exist, partly to do with the availability of coal nearby, but also with the level of industrialization of society (and vice versa).

Apart from a few scattered local uses of oil and gas, the large-scale use of oil only started late in the nineteenth century. The first commercial oil refinery factory was built by the Russians in 1837 in Baku, while the first wells in the US and Canada were dug in 1859 and 1902, respectively. Concurrent developments in the internal combustion engine (by now familiar names such as Daimler, Benz, Maybach) made oil a more attractive source of energy. Similarly, the development of electricity by Edison was a component of the big energy transition, where electricity provided a source that could be transported and delivered easily to the doorstep of its users. It was not just the invention of the lightbulb that made Edison famous—his vision of an electricity system with steam turbo generators and high voltage transformers is still very much the basis of our current system[17].

Around the end of the nineteenth century the first estimates of carbon emitted by the burning of fossil fuel started to appear as scientists started to realize that this growing source needed to be taken into account when studying the atmospheric concentration of CO$_2$. In Chapter 2 we have seen how Högbohm, Arrhenius, and Callendar made the first estimates of around 0.5 Gton C per year. Initially the interest was in explaining the role of CO$_2$ in the Ice Age, but soon Arrhenius and Calendar realized the importance of growing emissions on the global temperature. They assumed that all coal would be converted to CO$_2$ in the atmosphere, which is not correct. Later estimates were more careful and realized that there was an efficiency in the burning that needed to be considered as it depended on the type of coal[18]. Now the procedure is as follows: first one gathers industrial data on how much coal, brown coal, peat, and crude oil is extracted in a particular year in a particular country. Not all extracted oil is used in the country itself, as there is considerable trade in fossil fuels, so a country's net balance has to be drawn up that includes exports and imports. Luckily this type of data is generally easily available, as there are obvious economic interests attached to it. Sometimes, however, it is precisely the economic value that makes the data hard to obtain as

---

[17] See Smil, V., 2017
[18] Andrew, R., 2020

it becomes classified. Next, we need to convert the fuel into emissions. This depends on the type of fuel, the carbon content of coal and how much of that would be converted into CO_2.

Once these calculations are done it is possible for all countries to estimate their emissions due to fossil fuel burning (Figure 8.2). There are several things to note in Figure 8.2. First, up to around 1950 most of the emissions were caused by coal burning with oil and gas contributing mostly in the latter part of the twentieth century. Overall, the rise is approximately exponential, leading to a 2019 value of about 36.5 Gton CO_2 per year. The growth in these emissions decreased between 1960 and 2000, from 4.3% yr^{-1} for 1960–1969, to 3.1% yr^{-1} for 1970–1979, 1.6% yr^{-1} for 1980–1989, and to 0.9% yr^{-1} for 1990–1999. After this period, the growth rate began increasing again in the 2000s at an average growth

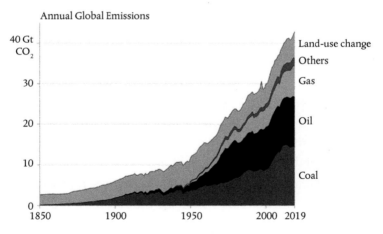

Figure 8.2 Increase in global CO_2 emissions in Gton CO_2 yr^{-1} from burning fossil fuel and partitioning between types of fuel. Emissions due to land-use change are also shown.
Data from the Global Carbon Project, 2021

rate of 3.0% yr^{-1}, but decreasing to 1.2% yr^{-1} for the last decade (2010–2019)[19]. A positive growth rate implies ever more emissions and ever more CO_2 remaining in the atmosphere. There was a sharp increase in the use of gas and oil after 1950, and a sudden increase in coal burning after 2000. In 1950 the share of oil in what is called the energy mix was about 20%. In 1968 oil (1.5 Gton C yr^{-1}) surpassed the amount of coal (1.45 Gton C yr^{-1}), but in 2005, due to the increased use of coal in China, coal (3.14 Gton C yr^{-1}) surpassed the use of oil (3.03 Gton C yr^{-1}). Gas use rose exponentially from a paltry 0.1 Gton C yr^{-1} in 1950, to 2.04 Gton C yr^{-1} in 2019. Most of the emissions from coal are stable or going down, but those from China and India have been going up since 2000. Emissions from oil in the EU and US over the last two decades have been declining slowly but are rising in other parts of the world (India, China). This is mainly due to the opening of the Chinese economy and the explosive growth of the Chinese and Indian economies around that time. Up to 1950 most of the emission came from Europe and the US (roughly 90%). That picture changes in the latter part of the twentieth century with China, the rest of Asia, and some countries in Latin America picking up. In 2006 China became the largest emitter, surpassing the US.

The early use of fossil fuel has led to a large share in the total emissions for the EU27 and USA, with a declining share for the UK where the first large industrial use of coal took place. The share of China, others (including Brazil), and India is rising, but the EU and US share the largest burden in the total amount of CO_2 put into the atmosphere in the past, namely: USA 25%, EU27 17%, China 13%, Russia 7%, UK 5%, Japan 4%, and India 3%. Over the last 30 years

---

[19] Friedlingstein, P., et al., 2020

this distribution changes to: China 21%, USA 19%, EU27 12%, Russia 6%, India 5%, Japan 4%, UK 2%. In a nutshell one sees here the problem of climate negotiations (see Chapter 10). Historically the western world, i.e. the USA and Europe, are responsible for over 50% of the cumulative emissions, while in the last decade the share of China, India, and other countries is rising. Who is going to commit to reductions? The group of countries with the largest historical share, or those with the highest current emissions? We will come back to these discussions in later chapters.

But what drives these emissions? Is it population growth, economic activity (or Gross Domestic Product), or other factors? It is possible to express all of that in a single equation, the so-called Kaya identity[20]. This states that the total emissions are a product of four driving factors: population growth, the world GDP, global primary energy consumption, and the carbon intensity. The latter are expressed in the Kaya identity as a ratio (F) against population (P), GDP (G) and energy (E): $F=P(G/P)\times(E/G)\times(F/E)$. The nice thing about the Kaya identity is that it allows us to investigate the drivers of fossil fuel growth, per region if we wish to. An analysis shows that GDP continues to grow worldwide; however, the energy gained by a unit GDP, the energy efficiency, has declined. The amount of CO_2 per unit energy has remained fairly stable. Overall, this leads to a 40% increase in energy use and fossil fuel emissions over the 30-year period from 1990. If one looks at the Kaya identity for China, one sees a decoupling between GDP and fossil fuel emissions, because the amount of energy used per GDP unit is slowly declining. In a country, say Pakistan, which is still relying on traditional fossil fuels, the coupling between GDP and

---

[20] Raupach, M., et al., 2007

emissions is very strong. To reduce emissions, the decoupling of GDP and fossil emissions is key.

The history of emissions over the last 50 years also shows another interesting phenomenon. After a crisis, such as the 2008 financial crisis, emissions temporarily reduced. This is the Kaya identity working: a reduction in GDP causes a decrease in emissions. However, the next year, the emissions are almost back at the same level (or similar growth rate to be precise). This has happened in both the first and second oil crisis, the dismantling of the Soviet Union, and has now also happened after the 2020/2021 COVID-19 crisis as well. After large reductions of up to 7% in emissions during the lockdowns, China started to accelerate their economic growth rapidly once the lockdown was lifted, with Europe and the US also showing signs of a quick recovery. At the time of writing, the world is holding its breath to see what the impact of economic recovery on fossil fuel emissions will be: have we been able to decouple growth more from emissions? The first signs are not encouraging.

Like the fate of the emissions from past land-use change, it is not immediately obvious how these emissions translate into changes in atmospheric concentrations. However, by analysing the gas bubbles in ice and since 1958 (Chapter 8) from direct measurements we have a good reconstruction of the past CO_2 concentration in the atmosphere. From this we know that on average about 45% of the fossil fuel emissions ends up in the atmosphere. When this was first discovered people started talking about a missing sink: where did the rest of the 55% go that did not go into the ocean? The oceanographer Revelle (Chapter 9) thought that the oceans would take up most, but further research has shown that both the ocean and the land take up the CO_2, the land biosphere ($\approx$30%) a little bit more than the ocean ($\approx$25%).

There are several ways to estimate the partitioning between the land and ocean, while the atmospheric concentrations are directly measured. One of the more elegant ways uses the fact that oxygen is produced by photosynthetic plants while CO_2 is taken up[21]. If one knows the minute decline in oxygen content of the atmosphere—after all we burn (oxidize) fossil fuels—from very precise measurements of the nitrogen to oxygen ratio (nitrogen making up 78%), it is possible to split this between land (where CO_2 is taken up to produce oxygen) and the ocean (where this does not happen in a similar way).

In Figure 8.3, however, models are used to reconstruct the biospheric uptake by the land and ocean. One of the most striking features of Figure 8.3 is that the uptake by the land and ocean appear to have grown in line with the increase in emissions. While each of these models may not be completely correct, their average or ensemble appears to give a very good indication of the uptake. The ratio of atmospheric increase over emissions stands on average at 45%, and has remained remarkably constant since 1958. While there have been small variations no definite trend has yet been established. This is of course good news! The ocean and biosphere act as a hoover taking up excess CO_2. It is, however, to be expected that at some point in time, this 45% will increase when the land saturates (trees cannot grow indefinitely), and decomposition of organic material is known to respond more quickly to changes in temperature than photosynthesis, shifting the balance from uptake (sink) to emission (source). This may happen within

---

[21] Keeling, R., Piper, S. and Heimann, M., 1996

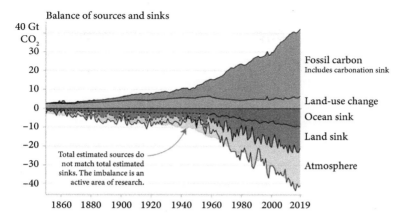

Figure 8.3 Annual CO_2 emissions from fossil fuel burning and cement production and land-use change sources (positive y-axis) and sinks in the ocean, land, and atmosphere (negative y-axis). The imbalance in the budget is also shown. Both in Gton CO_2 yr^{-1}.
From the Global Carbon Project, 2021

the next few decades when climate change induced droughts and nutrient limitations will cause the ecosystems to decrease their net CO_2 uptake. The first signs of this response are noticeable in the world's tropical forests[22]. Similar signs in the ocean are not visible: the ocean uptake still increases linearly with the increase in atmospheric CO_2 and is expected to continue doing so for some time. However, when large changes in ocean circulation occur, known as overturning, this may also change[23]. Decreases in the land and ocean uptake have immediate implications for the atmospheric CO_2 concentration that will inevitably go up if emissions do not decrease.

[22] Crisp, D., et al., 2021
[23] Crisp, D., et al., 2021

The final impact of the emissions from burning fossil fuel since 1850 on the atmospheric concentration in 2019 can be partitioned as follows. The contribution of coal was an estimated 99 ppm, oil 75 ppm, and gas 33 ppm. Cement production counts for an additional 5% (note that this is similar in magnitude to the 5 ppm we estimated for the effects of land-use change before 1850). Land-use change since 1850 contributed almost as much as coal burning (98 ppm). The land sink almost equals the land-use change emission (99 ppm), the ocean takes up a further 76 ppm. Calculating the budget and accounting for a small sink of cement carbonation yields an imbalance of about 9 ppm. The atmospheric concentration in 2019 was 410 ppm, the highest in the previous 3.5 million years with the increase completely caused by humans, and still increasing.

CHAPTER 9

DETERMINING ATMOSPHERIC AND OCEANIC CARBON DIOXIDE

W*e get acquainted with ways to measure the concentration (mole fraction) of CO_2 in the atmosphere. Charles David Keeling started the now famous Mauna Loa series of measurements in 1958 that is now the iconic symbol of human impact on the climate. He made direct measurements of the mole fraction, rather than the indirect measurements obtained by chemical titration that were common at the time. When Keeling analysed the results of the first few years of his measurements, he found a distinct seasonal cycle, with high values in the Northern Hemisphere winter and low values in summer. This was related to the fact that most land with vegetation, in fact most land, is in the Northern Hemisphere and plants were starting to take up CO_2 in the spring, thus lowering the concentrations. Because the atmosphere mixes so fast, this was even visible in the measurements at Mauna Loa. He also noticed that there was a small gradual increase per year in the baseline. Soon he related this to the impact of fossil fuel burning.*

In 1957 Roger Revelle and Hans Suess stated famously that 'human beings are now carrying out a large-scale geophysical experiment of a kind that *could not have happened in the past nor be reproduced in the future.' They investigated how much of the emitted CO_2 could be taken up by the ocean and, importantly, how much would eventually remain in the atmosphere to influence climate. Bert Bolin, later the first IPCC chair, refined their calculations and showed that uptake of CO_2 by the surface ocean is one thing, but bringing it down to the deep ocean quite another. The ocean plays a key role in the carbon cycle. To investigate this, a new research programme was established called Geochemical Ocean Sections (GEOSECS). This programme was to measure systematically the chemical composition of the ocean. The GEOSECS observations, originally aimed at better understanding the transport of water masses, were hugely important in the further development of chemical oceanography and in particular ocean carbon cycle science. Measuring carbon in the ocean is much more difficult than in the atmosphere as carbon in the ocean is distributed less homogeneously and comes in three different forms, dissolved CO_2, carbonate, and bicarbonate. The three are in intricate chemical equilibrium, which can shift depending on the availability of one or more of the components. This equilibrium forms the basis for the buffering capacity of the ocean, an issue that Revelle and Suess had raised in their 1957 paper and that is now called the Revelle factor, indicating how efficient the ocean is in taking up CO_2.*

The US-based Japanese scientist Taro Takahashi meanwhile did for the ocean measurements what Keeling had done for the atmosphere—he developed a system to directly measure the concentration of CO_2 in ocean surface waters. These observations were followed up by the World Ocean Circulation Experiment and Joint Global Ocean Flux Study (in the late 1980s and 1990s). The basis was laid for a programme of ocean observations. In the atmosphere the coordination was taken up by the WMO's Global Atmospheric Watch programme, that built successfully on the initial US networks set up by Keeling and NOAA. In the ocean the SOCAT programme collects *and coordinates surface observations and the GLODAP programme those in the deeper ocean.*

On land a larger network of stations now regularly collects concentration measurements. Recent developments in satellite technology hold promise that soon we will be able to observe carbon dioxide from space in great detail and determine sources and sinks more precisely.

The iconic image of human influence on our environment today is the sheer relentless increase in the amount of carbon dioxide or CO_2 in the atmosphere. The measurements on which this image is based were started by Charles David Keeling slightly more than 60 years ago. In 1953, Charles Keeling, with a freshly acquired PhD in polymer chemistry from US Northwestern University in his pocket, decided not to embark on the traditional career in the chemical or oil industry that his study had prepared him for. Instead, he opted for an academic position in geochemistry because he had become increasingly fascinated by geology. He chose to start his further career at the California Institute of Technology in Pasadena, with Professor Harrison Brown who had just set up a new department to study geochemistry: the chemistry of the Earth. In doing so Keeling would become the first scientist who would unequivocally reveal the impacts of burning fossil fuel on the composition of our atmosphere.

In California he developed the experimental set-up that to a large extent would form the base of both his own scientific career[1] and of a future observing network for carbon dioxide in the atmosphere. His supervisor was interested in how carbonate from

---

[1] Keeling, C., 1998

limestone, dissolved in river waters, would eventually achieve equilibrium with the CO_2 in the air. To measure this, Keeling developed a system whereby he could spray water onto a glass surface in a confined space and let the water come into equilibrium with the air. A modified version of this equilibrator is still the main tool to measure the CO_2 in surface ocean waters all over the world. He subsequently needed only to measure the CO_2 content of the air in the confined space to know the concentration in the water.

These concentration measurements of CO_2 in water were conventionally done using chemical titration methods that were rather variable and unreliable. One needed, for instance, to be sure that all of the CO_2 was taken out, a tough process that easily led to underestimation of the CO_2 content of the sample.[2] Keeling was not satisfied with this situation, and he developed a novel way to determine the concentration, based on measuring the actual pressure CO_2 exerted. He found a device in the literature called a constant-volume manometer (Figure 9.1) and with this instrument he was able to achieve an accuracy of one thousandth of the measurement (so for a concentration of 300 ppm,[3] the error or imprecision would be ±0.3 ppm). After he had sampled the air in a vacuum flask of known volume, he froze and extracted the CO_2 gas into a cold trap that is connected with the manometer device using liquid nitrogen. After warming up the frozen CO_2 in the thermally stabilized manometer device of well-known volume, he observed the change in pressure resulting from the gas and

[2] Siegenthaler, U., 1984
[3] ppm is the unit that expresses the quantity of a gas in the number of particles of that gas per million other particles on a volume basis. In this case, 300 ppm implies 300 molecules of CO_2 in a sample of a million other molecules. Note that this is a mole fraction. Mole fractions are generally measured in dry air.

Figure 9.1 Charles David Keeling and the manometer he designed, holding a beaker with liquid nitrogen. The manometer allowed him to measure the atmospheric concentration of CO_2 with unprecedented precision.
Photo credit: Scripps Institution of Oceanography

could then determine its amount. The manometer method was a major breakthrough as no instrument with that precision to measure CO_2 existed at that time.

To solve the initial question of how the limestone waters equilibrated with the CO_2 in the air he also needed an estimate of the value of its atmospheric concentration. Writing 60 years later, when every newspaper seems to mention this value almost once a week, is seems strange that such a value did not exist. Yes, textbooks mentioned values of about 0.3%, but no more precise value was provided, although the Englishman Guy Callendar[4] had collected existing values of 292 ppm from earlier measurements around 1900 and 312 ppm around 1930 (Chapter 2). So, Keeling started taking samples of the air, first in the city (but these turned out to be highly variable due to human influence), and later in the pristine air of Big Sur State Park close to the

[4] Callendar, G., 1938

Pacific Ocean. The measurements all came out at around 310 ppm, with values obtained in the night somewhat higher than those during the daytime. Taking air samples at different locations, as far away as Canada and later with samples collected by colleagues on ocean vessels and aircraft, Keeling was able to conclude that the concentration of CO_2 in the atmosphere in the Northern Hemisphere was spatially remarkably constant at about 310 ppm.

In July 1957 the International Geophysical Year (IGY) had finally started. Conceived in 1950, the preparation had taken considerable time. IGY called for an unprecedented global effort at improving the understanding of the Earth System. It was also meant to improve relations between the Soviet Union and western countries, where the Cold War was hindering scientific collaboration and progress. The IGY was coordinated by the International Council of Scientific Unions with the World Meteorological Organization and many weather bureaus worldwide participating. Among many other areas such as polar science, towards which a considerable research effort was directed, the IGY would also have implications for CO_2 observations.

In preparation for the IGY, the Swedish meteorologist Carl Gustav Rossby (perhaps better known for his contribution to the planetary scale waves encircling the planet, called Rossby waves) was also interested in measuring CO_2. His co-worker, Stig Fonselius, used barium hydroxide to bind the CO_2 and then titration methods with hydrochloric acid to determine the remaining amount of barium hydroxide, one of the chemical methods that had made Keeling look towards an alternative. The Swedes were primarily interested in using CO_2 as a passive tracer to track the movement of airmasses.[5] Unfortunately, their measurements turned out to be rather inaccurate, varying from 319 to 347 ppm between locations.[6] They attributed this to the movement of the air masses; but, as we have seen from Keeling's measurements, generally the background concentration of CO_2 does not change that much between locations.[7] They did, however, appreciate the value of setting up a global network of stations that employed the same technique everywhere: a procedure that minimizes systematic errors: *'If we are to detect trends, we need samples taken by the same technique, at the same hour of the day and at the same place for several years, and the analyses have to be carried out with identical techniques. The best thing would be to get samples from a network covering a representative portion of the earth's surface. This ambitious programme we have not yet been able to accomplish.'*[8]

The IGY provided the stimulus for the US National Science Foundation to award funds to Keeling to initiate such a network, initially of four stations around the world, each measuring CO_2 concentration using new devices, to be calibrated by his accurate manometer. One of those locations was the US Weather Bureau's monitoring station on top of a Hawaiian volcano, Mauna Loa.[9] Another was to be deployed at the Little America research base in Antarctica, yet another one was reserved for deployment on a ship, and the last one remained at CalTech for comparison

---

[5] Fonselius, S., Koroleff. F, & Warme, K., 1956
[6] Note that Keeling, in his paper from 1998, claims a range of 150–450 ppm, which cannot be reconstructed from the Fonselius, et al., 1956 paper.
[7] It is interesting to keep this idea in the back of our mind, as air masses contain information in their CO_2 content, forming the basis of so-called inverse modelling techniques that use known atmospheric transport to back-calculate the origin of the CO_2.
[8] Fonselius, S., Koroleff. F, & Warme, K., 1956
[9] Mauna Loa is a volcano, so a large amount of correction must be applied to the measurements to take out those that stem from volcanic emissions.

purposes. The Weather Bureau was the forerunner of the National Ocean and Atmospheric Administration (NOAA)—it would play an increasingly important role in the measurements. Not only was the Weather Bureau's Harry Wexler crucial in getting Keeling's initial programme going, but NOAA also provided support for the Mauna Loa measurements. Keeling wanted to deploy new devices that were called Non-Dispersive InfraRed analysers, which used the fact that CO_2 absorbs in the infrared region of the radiation spectrum (Chapter 2). If you have a source emitting long-wave radiation in the infrared region, like a hot coil of wire, and a detector, with air flowing in between, the amount of radiation received by the detector indicates how much is absorbed on its way and thus how much CO_2 is in the air. In a way it was the reverse of the experiments performed by Tyndall (Chapter 2). The amount of water vapour complicates the measurement, as we know it also absorbs radiation in the infrared region. Therefore, in today's measurement systems water is usually taken out by passing the air through a cold trap or a chemical scrubber. But the principle is sound and based on solid physics. The big advantage of these instruments was that they could provide a continuous measurement, as opposed to single event sampling with, for instance, large flasks that had to be returned to the lab for analysis. The disadvantage was that the accuracy and stability of the instrument was less than the manometer device Keeling had developed earlier, but by regularly calibrating the instruments, this disadvantage could to a large extent be overcome.

Enter Keeling's boss, Roger Revelle. He was a renowned oceanographer, and an apparently somewhat pompous director of Scripps Oceanographic Institute. In 1957 he wrote a seminal paper with Hans Suess in which they suggested that the ocean had taken up most of the carbon released by fossil fuel combustion. This suggestion was not new; it had been raised as early as 1903 by Svante Arrhenius. They did go on, however, to suggest that '*it therefore becomes of prime importance to attempt to determine the way in which carbon dioxide is partitioned between the atmosphere, the oceans, the biosphere and the lithosphere.*'[10] They approached the question of where CO_2 was taken up, if at all, by using the radioactive isotope of carbon, carbon 14 or ^{14}C (Chapter 5). When the Austrian scientist Hans Suess visited Libby's lab at the University of Chicago in the aftermath of the Second World War, he quickly saw the potential for applying the radiocarbon technique to environmental problems,[11]

Hans Suess, working then for the US Geological Survey, had determined what is now called the Suess effect by analysing the ^{14}C content of tree rings. The Suess effect arises from the fact that when we burn fossil fuel, the fuel is geologically very old (usually several hundreds of million years). This implies that any ^{14}C that was in the atmosphere and taken up by vegetation through photosynthesis at that time, would have lost its radioactivity. If that fossil fuel is burned and converted to CO_2, this will then tend to decrease the $^{14}C/C$ ratio of the atmospheric CO_2. In two papers[12] Suess analysed various materials, among them the wood from two trees from the east coast of the US and one from Alaska. He initially noted that the dilution effect amounted to 3.4% over a period of 50 years, with the Alaskan tree showing lower values. He subsequently revised this value to 1% and later to 1.7% to account for local pollution effects.

[10] Revelle and Sues, 1957
[11] Waenke, H. & Arnold, J., 2006
[12] Suess, H., 1955 and Suess, H., 1954.

In 1955 Roger Revelle had convinced Hans Suess to join Scripps and his rapidly developing group of (ocean) geochemists. Suess and Revelle quickly grasped the potential for applying this natural tracer to more recent climate history when they teamed up to produce their landmark 1957 paper. In it they famously stated that '...*human beings are now carrying out a large-scale geophysical experiment of a kind that could not have happened in the past nor be reproduced in the future*'.[13] One of the critical questions in establishing how much of the CO_2 originating from the human burning of fossil fuels remained in the atmosphere concerned the role of the ocean. If it took up CO_2 fast, that was one thing, but it also needed to be transported into the deep ocean, and nobody at that time had a clear idea how much time that would take, with estimates varying from several decades to several tens of thousands of years. Radiocarbon estimates would in principle be able to help address that question since they would provide indications of the age of the waters. By comparing the ^{14}C age of marine material against that of wood of terrestrial origin, they concluded that the average exchange time of a CO_2 molecule in the air before it was absorbed by the ocean was about 10 years: '*We conclude that the exchange time $\tau(atm) = 1/k_1$, defined as the time it takes on the average for a CO_2 molecule as a member of the atmospheric carbon reservoir to be absorbed by the sea, is of the order of magnitude of 10 years.*' Now things got a bit more complicated, as Revelle and Suess noted that not all the CO_2 that is present in the atmosphere, in particular that fraction arising from fossil fuel emissions, gets absorbed at the same rate with the same residence time. An increase in the atmospheric CO_2 does not simply lead to a comparable increase in the carbon content of the surface ocean. This is because the ocean carbonate system

[13] Revelle, R., & Suess, H., 1957

acts as a buffer. We will go deeper into this phenomenon later in this chapter when we discuss the first ocean observations. Suffice it here to say that the net result is that the increase in surface ocean carbon content is only 10% of the increase in atmospheric CO_2. Some authors think that Revelle added this effect only as an afterthought to the paper.[14] While it indeed reads a bit like that in the paper, the uptake reduction to only 10% is now known as the Revelle factor. With that reduction in uptake capacity, Revelle and Suess concluded that it was *'quite improbable'* that the increase in the atmospheric concentration of CO_2 could become higher than 10% of the fossil fuel emissions at that time. They still believed that the ocean, a carbon reservoir about 50 times larger than the atmosphere, was absorbing most of the additional carbon dioxide. However, by extrapolating the fossil fuel use in time, they estimated that the atmospheric concentration could eventually rise by about 20–40% over the next 50 years. The uncertainty in their estimate made them call for the IGY to start measurements on the 'geophysical experiment' of adding CO_2 to the air.

It was down to the Swedes Bert Bolin, the successor to Karl Rossby at the International Meteorological Institute in Stockholm, and Erik Erikson to put more substance to Revelle and Suess's afterthought. In a careful and thought-through analysis they concluded that an increase of about 10% in atmospheric CO_2 might well be compatible with a Suess effect of a radiocarbon reduction of a few per cent, as observed.[15] They showed that the absorption capacity of the ocean could well be overestimated by previous accounts, largely as a result of the slow mixing of CO_2 to the deeper layers of the ocean. While surface uptake is the

[14] Archer, R. & Pierrehumbert, R., 2011
[15] Bolin, B. & Erikson, E., 1959

so-called rate-limiting step, the longer-term uptake depends, of course, on how fast the surface carbon can be transported to the deeper ocean. This is a much slower process and much harder to determine, as the increase in anthropogenic carbon in the ocean is small against the large background of carbon overall in the ocean. Bolin and Erikson further suggested that the amount of CO_2 that remained in the atmosphere would be much more dependent on the absorption of the biosphere, the land. This proved a remarkable foresight, to which we will return in the next chapters. They also suggested that, with an increase in atmospheric CO_2 of somewhere between 0.1 and 0.3% per year, *'It should therefore be possible within a few years to observe whether an increase occurs with this computed rate or not.'*[16] These were the measurements for which Revelle had just brought in Charles Keeling, and for which unfortunately the measurements of Fonselius had been too inaccurate.

These new measurements were not initially meant to provide a long-term monitoring programme over many years that would be able to detect so-called interannual variation, the year-to-year variability in atmospheric CO_2 concentration. Rather, in Revelle's view they were aimed at providing a snapshot of atmospheric CO_2 concentrations, that would have to be repeated after 20 years to see if there really had been an impact of fossil fuel emissions on atmospheric CO_2. To obtain a good snapshot estimate, Keeling's new instruments would thus also be deployed on-board an aircraft and a ship.

It is worth quoting at some length Keeling's own words describing what he saw when the first full year of measurements of the Mauna Loa station became available (see Figure 9.2): '*A regular*

---

[16] Bolin, B. & Erikson, E., 1959

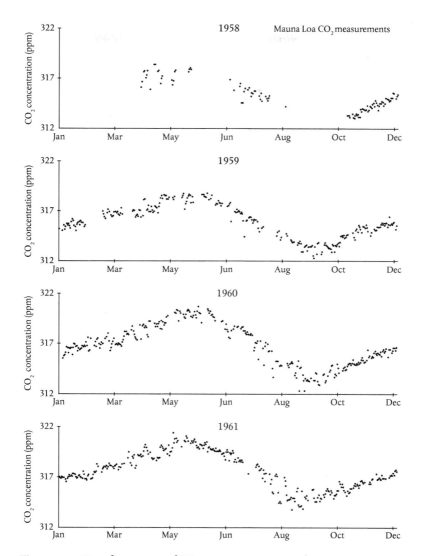

Figure 9.2 First four years of CO_2 measurements at the Mauna Loa station.

Redrawn after Pales, J. & Keeling, C., 1965, with revised data after Walker, S., Keeling, R. and Piper, S., 2016

*seasonal pattern began to emerge, but it differed markedly from earlier pub-
lished northern hemisphere data in which the maximum concentration was
typically in January, a month when CO_2 from burning is likely to accumulate in the air near the ground because of temperature inversions....
The maximum concentration at Mauna Loa occurred just before plants in
temperate and boreal regions put on new leaves. At Mauna Loa the regular
seasonal pattern almost exactly repeated itself during the second year of
measurements .... We were witnessing for the first time nature's withdraw-
ing CO_2 from the air for plant growth during the summer and returning it
each succeeding winter.*[17]

What Keeling observed was the breathing of the planet with
plants taking up atmospheric carbon in spring and summer and
returning it to the atmosphere by decomposition in the autumn
and winter. The first part of the puzzle was laid: annual mean
atmospheric CO_2 concentrations varied globally little within a
year, but did show a pronounced seasonal cycle.

In 1961, Bert Bolin invited Keeling to spend a year at Stockholm. Keeling used the year to analyse his initial results. Bolin
and Keeling looked in particular at the variation by latitude and
discovered that the amplitude of the seasonal variation in the
CO_2 concentration was lower in the Southern Hemisphere than
in the Northern Hemisphere. The seasonal cycle was weaker in
the Southern Hemisphere. It became clear that natural sources
and sinks together within ocean and land masses were inter-
acting to produce these spatial gradients. Precisely at a point
where more data over more years would have helped to solve
some of these issues, the programme ran into trouble. Cuts in US
Weather Bureau funding severely reduced staff numbers at Mauna

[17] Keeling, C.D., 1998

Loa, while measurements on aircraft and ships, and even at the South Pole, were forced to stop. Monitoring the environment for long-term purposes was not high on the agenda of funding agencies.[18] The measurements at Mauna Loa were luckily restored after Keeling secured new funding from the NSF. His battle with the funding agencies to keep his measurements alive was however only starting.[19]

Revelle and Suess in the meantime were aiming to understand how much CO_2 would be taken up by the ocean. This is not as simple a problem as it sounds. Mixing and circulation in the atmosphere is relatively fast and generally acts to smooth the distribution of any long-lived chemical species. Hence the remarkable constancy of the global atmospheric mole fractions of CO_2 that Keeling measured. In contrast, mixing in the ocean is considerably slower and natural physical and biological processes create a strong variability in the concentrations, both in space and time. As a result, the distribution of anthropogenic CO_2 and, for that matter, other tracers in the ocean becomes non-uniform. This introduces a serious sampling challenge if we want to determine changes in the amount and distribution of ocean carbon. While in the atmosphere, one atmospheric station, say Mauna Loa, may be sufficient to determine the annual mean concentration of CO_2, and the long-term trend, provided the measurement error is small; in the ocean we need many more measurements. Inevitably, these are limited in space and time, because they have to be taken on-board ships or more recently from robotic vessels.

---

[18] In fact, it is still not very high on the agenda of funding agencies, who generally prefer short term 3-4 funding cycles of hypothesis-driven research as opposed to long-term monitoring, although this situation is slowly improving.
[19] Keeling, C.D., 1998

A further complicating factor is that while the surface exchange of CO_2 between the ocean and the atmosphere is fast and substantial, storage and further transport of carbon goes through the deeper layers of the ocean where mixing takes several centuries or even thousands of years. All in all, no easy matter.

Oceanographers, among them the famous Henry Stommel, had been looking at using geochemical tracers to study deep ocean circulation. They called for an ocean observing programme to study the circulation in greater detail at around the same time as the atmospheric observations had started. The following transcript of a conversation illuminates how such programmes originate, once they find suitable ears and an environment in which to mature. '*Ed Goldberg [Scripps Institution of Oceanography] and I [Wallace Broecker] were attending some sort of meeting at WHOI [Woods Hole Oceanographic Institution] during the late 1960s. Hank (Stommel) came to us and said that radiocarbon measurements in the sea were of great importance. He went on to gently chastise us [the geochem community] for doing only scattered stations. What is needed, he said, is a line of stations extending the length of the Atlantic. Gee, we said, that would cost a million dollars, a sum greater than the entire NSF annual budget for ocean chemistry. Hank replied, "Well it would be worth more than a million." He spurred us to propose such a venture. Soon plans were being formulated not only to do carbon-14 but also a host of other chemical and isotopic properties along Hank's Atlantic line.*'[20]

The programme Stommel wished for so keenly was to be GEOSECS, short for Geochemical Ocean Sections. The GEOSECS programme was constructed around three key principles. First, the necessity of a sampling network of high density in both the

---

[20] Farrington, J., 2000

vertical and horizontal, to be able to sample the heterogeneity between surface and deep ocean and within basins as well. Second, a considerable effort was planned to study particulate transport (transport of particles) of the various tracers that were thought to remove some of the tracers directly to the sediment, rather than by diffusion. Third, an effort should be made to sample as many species and constituents as possible, in particular those that showed strong vertical and horizontal variations in the sea. The programme was part of the International Decade of Ocean Exploration and included expeditions by the United States, France, West Germany, and Japan, amongst others. Figure 9.3 shows the proposed North–South transects in the Atlantic and Pacific that were part of the original US contribution. At each stop/station vertical profiles of 50 samples, each measuring about 30 litres of

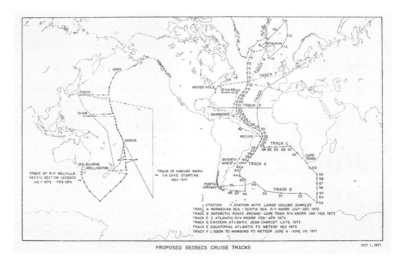

Figure 9.3 Proposed GEOSECS Atlantic and Pacific sampling lines. These are also known as the GEOSECS I and GEOSECS II transects.

ocean water, would be taken while, at other designated stations, very large samples of 270–1000 litres were proposed to be taken to measure trace constituents and low-concentration radioisotopes (such as ^{14}C).

The samples would initially be analysed on-board, but with some being taken back to land for further analysis. The GEOSECS transects were hugely important in the further development of chemical oceanography and in particular ocean carbon cycle science. To understand the distribution of ^{14}C, one of the important tracers the distribution of which they wanted to research, the full carbon budget of the ocean needed to be studied. In fact, our first understanding of carbon cycling in the ocean came in through the back door of a programme aimed at understanding the transport of water masses in the ocean. Not that everything went smoothly. When the research vessel *Knorr* left Woods Hole for a nine-month cruise on 18 July 1972 it lost the entire conductivity-temperature-depth probe/profiler (CTD)-rosette sampling device on the first station they stopped at. Apparently, a locking pin had not been put into its place properly and the very expensive device ended up on the bottom of the ocean. It would take months to assemble a new one.

In contrast to the atmosphere, carbon in the ocean is distributed less homogenously and comes in three different forms, dissolved CO_2, carbonate (CO_3^{2-}) and bicarbonate (HCO_3^-). A negligible contribution is made by a species called true carbonic acid (H_2CO_3). At a pH of 8.2, the majority of carbon in the ocean consists of bicarbonate (89%), the remainder being made up of carbonate (10.5%) and CO_2 at only 0.5%. Most of the carbon in the ocean is thus not in the form of dissolved CO_2 gas. In the atmosphere, CO_2 is virtually inert. Things are rather different in the oceans. Here, an important complication arises when we realize that CO_2, carbonates, and bicarbonates are in an intricate chemical equilibrium with each other. Let us write down this carbon equilibrium (or carbonate system) in our shorthand chemical notation. First the one where CO_2 reacts with water to form bicarbonate and a proton: $CO_2 + H_2O \rightleftarrows HCO_3^- + H^+$. A second equation converts the bicarbonate ion into a carbonate ion, with again a proton: $HCO_3^- \rightleftarrows CO_3^{2-} + H^+$. The arrow between the two sides of the reaction goes both ways. This is important, as it effectively means that, if the oceans become more acidic, CO_2 is less easily absorbed: with large amounts of additional protons, the reactions would shift towards the left. This process is the basis of the buffer factor Revelle and Suess had been discussing in their 1957 paper.

This equilibrium determines the ratio of the carbon species, and thus also the balance of the charges of the carbonate system. Marine geochemists have developed the term alkalinity to deal with the availability of protons in seawater, sometimes also called the buffering capacity. A formal definition reads as follows: 'the total alkalinity of a seawater sample is defined as the number of moles of hydrogen ion equivalent to the excess of proton acceptors (bases formed from weak acids…) over proton donors (acids) …'[21] Most of the alkalinity in the ocean is determined by the carbonate alkalinity; only 4% of the seawater alkalinity is produced by other forms such as borate $(B(OH)_4^-)$. Alkalinity can be determined by titration, and this is the classic way of determining the balance between the three carbon species. In seawater we also need to determine the other components. It is possible to

---

[21] Dickson, A., Afghan, J., & Anderson, G., 2003

measure four parameters directly when we analyse a sample: the CO_2 partial pressure (pCO_2 in an equilibrator), the total dissolved inorganic carbon concentration (by adding acid of one form or other to the sample so that all the carbon species convert to CO_2), the acidity (or pH), and the total alkalinity. These parameters fully determine the carbonate system in the ocean. However, and luckily, knowing only two allows us to calculate all the parameters for given temperature, pressure, and salinity, once we know the exchange coefficients between the different species.

By around 1979, about 6000 determinations of alkalinity had been made in the GEOSECS programme. A meeting was then organized in La Jolla to determine the cause of a nasty but persistent difference between pCO_2 values that were directly measured and those determined through alkalinity titrations. The difference could run up to 15–20% for the Indian Ocean and Pacific, with the values from titration always being on the high side; researchers had struggled to find the cause of this difference. The mistake finally appeared to be in the way the total dissolved inorganic carbon concentration was determined by the titration method.

The direct pCO_2 measurements that highlighted the problem were obtained by Taro Takahashi. Takahashi, a Japanese immigrant to the US, had obtained his PhD in 1957 from Columbia University and went on to do a postdoc at Lamont Doherty Observatory, then known as the Lamont Geological Observatory of Columbia University, Palisades, New York. The job at Lamont was offered to him by the director, Maurice Ewing, standing next to him in the toilet, one day just after he graduated: '*And I was assigned to work with Larry Kulp. And Kulp was told by Ewing that hey, this kid is coming. So that's the way I became sort of an, you know, oceanographer, oceanographic geochemist.'*[22] His assignment, part of the original IGY studies funded by NSF that also supported Keeling's measurements, was to study carbon dioxide in the ocean and in the atmosphere to determine whether the ocean was in equilibrium or was taking up carbon dioxide from the atmosphere, the problem Revelle and Suess had addressed earlier. After an initial postdoc at Lamont, he spent eighteen months at Scripps Oceanographic Institution to work on similar problems under Roger Revelle and with Keeling. After these two postdocs his career took a different direction for the next 12 years or so while he studied high-pressure rock systems, before he returned back to study ocean carbon dioxide interactions. In 2010 he was awarded the UN Environmental Programme's Champion of the Earth award. Today he is best known for his global maps of surface ocean–atmosphere exchange of CO_2

In 1957, the year the IGY started, Takahashi boarded the Lamont research vessel *Vema* that would bring him from New York to the southern tip of Argentina and further on to Cape Town, South Africa. The *Vema* was a three-masted schooner and converted luxury yacht that was bought by Maurice Ewing, the same person that offered Takahashi the job, to serve as a research vessel for Lamont. Takahashi concluded after he had analysed the initial 470 hours of measurements, that *'the Atlantic surface water does not appear to be in equilibrium with the atmosphere above, as far as carbon dioxide is concerned. The partial pressures of carbon dioxide in surface ocean water exceeded those in the air between the equator and 40 °S. The opposite relation exists north of the equator and south of 40 °S during the summer months of the southern hemisphere. This indicates that the ocean water is releasing carbon dioxide*

[22] Takahashi, T., 1997

into the atmosphere between the equator and 40 °S, whereas it is absorbing carbon dioxide from the atmosphere north of the equator and south of 40 °S.'[23] He had identified the main pattern of ocean–atmosphere exchange of CO_2, with the tropical oceans outgassing and the Northern and Southern Oceans taking up.

The measurement system Takahashi developed for GEOSECS was in essence a refinement of the one he used 12 years before. It consisted of a gas-water equilibrator that used a rotating paddle wheel to stir the water (5 litres of a sample) in the equilibrator more strongly than using a spray as was done by Keeling in his initial experiments (see Figure 9.4).

The prime advantage of this system and fundamental to operating it on-board a ship was that both the marine air inlet (at the left of Figure 9.5) and the outlet were open, minimizing the differences in pressure. Once the air was in equilibrium the air would be moved to an infrared gas analyser to measure its mole fraction. The standards provided the calibration for the system. Takahashi claimed to be able to achieve an accuracy and precision of ±2 ppm.

In analogy to the fixed atmospheric stations over land (and ocean), the ocean observing network also initiated measurements at some fixed locations where continuous time series were produced at various depths, such as the Bermuda Atlantic Time Series (BATS) and Hawaii Ocean HOT station. These stations provided and still provide data from a large number of variables, at one location. HOT is in the Pacific Ocean at 22 ° 45′ N and 158 ° 00′ W, while the Bermuda Atlantic Time Series station is located in the North Atlantic Ocean at 31 ° 40′ N and 64 ° 10′ W. Measurements

---

[23] Takahashi, T., 1961

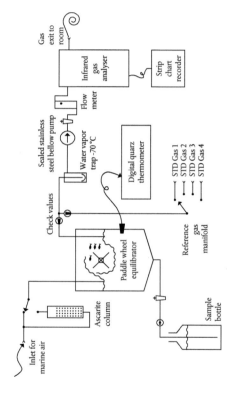

Figure 9.4 Schematic diagram of the dynamic CO_2 equilibration and measurement system developed by Takahashi for GEOSECS.

After Takahashi, T., Kaiteris, P. & Broecker, W., 1976

at both sites started in 1988 as part of the JGOFS (Joint Global Ocean Flux Study) programme; a research vessel visits the exact location once a month to take the measurements.

Based on the success of the programme a sequence of other scientific expeditions was planned and executed in the years following GEOSECS, such as the Transient Tracers in the Ocean (TTO) expedition of the North and tropical Atlantic Oceans in the early 1980s, and the World Ocean Circulation Experiment (WOCE) and Joint Global Ocean Flux Study (JGOFS) in the late 1980s and 1990s. The groundwork for a global observing system for ocean carbon was there.

The development of such a full observing system reflects the work done by many men and women who often suffered sometimes terrible difficulties and hardships on their way to getting the data. Not only technical or scientific problems needed to be solved, but funding needed to be secured, administrations to be convinced. Once that was organized, the data had to be taken, sometimes under extremely harsh conditions, particularly at sea or in places such as Antarctica. It is thanks to these unsung heroes, often appearing just as co-authors or barely mentioned in the acknowledgements at the end of papers, that the stories in this chapter can be told. Their persistence, perseverance, and dedication to science have made the developments told here possible. Thanks to them there is now, 60 years later, the outline of a global observing system that may in the near future be able to track atmospheric CO_2 and determine whether the values of the atmospheric and oceanic concentrations of CO_2 really correspond to the emission reductions that are promised in the international agreements. Let us now see how such an observing system developed from the initial measurements.

The IGY had provided the initial stimulus for the World Meteorological Organization to get its act together and to set up a network for the measurements of ozone. Not much later, a network of air chemistry stations was set in in place under the name BAPMon, short for Background Air Pollution Monitoring Network. This network initially provided measurements of precipitation chemistry only, then aerosol and, in 1975, carbon dioxide measurements were added. The addition of CO_2 to the measurement suite was likely the result of a 1969 report written by Christian Junge from Mainz, Germany and Charles Keeling with the recommendation to establish a worldwide CO_2 monitoring programme. WMO coordinated the network; they did not fund the stations—this was and still remains part of the individual nations' own obligations. This distributed funding model caused individual scientists considerable efforts and sometimes much pain when proposals for funding were rejected. As an example: while Keeling managed to secure funding in 1971 for a large contribution to the new WMO programme, in 1972 the winds at the National Science Foundation started to blow from a different direction. His research was not 'basic' enough and his budget was reduced by almost 50%. The next several years saw a continuous battle to secure funding, with now the Department of Energy also stepping in and the newly established National Ocean and Atmospheric Administration entering onto Keeling's grounds to help fund part of the initial programme.

Slowly now, however, a larger network was beginning to emerge. In 1968, the Weather Bureau's successor, the National Oceanographic and Atmospheric Administration (NOAA), started the Global Monitoring for Climatic Change programme. Lester Machta made his first measurements of CO_2 in 1968 at Niwot Ridge about

30 km straight west of Boulder in the Rocky Mountains, at high altitude on the continental divide.[24] At Mauna Loa in 1974, NOAA established a second measurement series, also under the direction of Lester Machta, to run in parallel to Keeling's measurements. They also continued to fund the operation of Keeling's monitoring at Mauna Loa. In 1976, new sites were established on Samoa and at the South Pole that were previously sampled by Keeling through flasks. All these new sites ran continuous analysers, supplemented by frequent air samples taken in flasks that were sent to Boulder for analysis as a back-up and calibration. This flask sampling was also a cheap and efficient way to sample remote sites such as the Falkland Islands, Guam, and Amsterdam Island. In the words of Lester Machta: 'We got the four stations and we needed more, and what I did was say, I can't get a lot of money, let's get volunteers in very, very remote locations and start a flask programme. We would ask people at no cost to us to fill up flasks and send them back to a central laboratory, where by this time we had set up automated equipment for carbon dioxide that could handle a flask very easily.'[25]

After the, mainly, US boost in network capabilities, the network was also starting to expand in other countries too. In New Zealand, Dave Lowe started regular CO_2 measurements at Baring Head in 1970; Pettit in the Canadian Arctic in 1975, Pearman in Australia in 1976 and Ciattaglia on Monte Cimone, Italy, in 1978. In 1976 the Cape Grim Baseline Air Pollution Station was established at the north-western point of Tasmania. In Germany, in 1972 the first measurements were made at Schauinsland. Located at 1200 metres above sea-level in the Black Forest in Southwestern Germany, this was in contrast to the other sites not a background

[24] Tans, P., 2019
[25] Machta, L., 1991

Figure 9.5 The Schauinsland observatory. The earliest measurements at Schauinsland started in 1965. Work at Schauinsland is focused on the detection of long-term trends. Schauinsland hosts, since 1972, the longest continuous observation of carbon dioxide (CO_2) observations in Europe.
Reproduced with permission of the photographer, Frank Meinhardt.

station, but one of the first continuously measuring stations in a continental area. The Schauinsland station was established as early as 1967 as part of the BAPMon network, and measurements of CO_2 started in 1972.

While making measurements at a single location, and analysing the sample in a single lab with an unchanging technique and method is one thing, it is quite another to make sure that measurements of the CO_2 mole fractions made by a different institution are comparable to the ones, made for instance by Keeling in his Scripps Lab. With the network expanding this became a critical issue early on and it was decided that a Central Calibration Lab was to provide standard gases against which the calibration gases used at other laboratories in the world could be calibrated. This task comes down to maintaining a set of primary standards that have known mole fractions (concentrations) of CO_2 still regularly determined with a manometric system, similar to the one Keeling used in his early times. From these, secondary standards are derived that further give rise to standards that can be used to calibrate the standard gases from different labs. The development and production and certainly the maintenance of the primary standards is key to the success of a network. On the recommendation of WMO experts, this task was initially performed by Keeling's laboratory at Scripps with funding from the UN Environment Programme but was taken over by NOAA in 1995 after a long battle with the US National Bureau of Standards who claimed that they were also able to produce the primary gas standards. There is an interesting parallel story about the need to improve reliable measurements of carbon in seawaters using reference standards, again produced by Scripps.[26] With these it became possible to compare anthropogenic signals in the ocean between different cruises.

The initial idea to involve the Bureau of Standards came from NOAA's Lester Machta, who thought that enlisting the US top-level agency on metrology standards was the key to providing the long-term sustainable and maintainable standards that were needed. The big concern was, that, if inadequately kept, the standards would deteriorate, as small amounts of CO_2 would eventually be fixed by the stainless-steel interiors of the high-pressure cylinders and deteriorate the standard. The standards the NBS produced, however, were slightly different from the ones Keeling was providing and discussion arose about which one to use. While

[26] Dickson, A., Afghan, J., & Anderson, G., 2003.

the community asked for standards at 0.1 ppm accuracy, the standards bureau was opting for a lower 0.3 ppm. They never really achieved the level of precision required and were not capable of providing the required constancy of the standards the community felt was needed. After several highly technical and politicized discussions in the WMO working group in 1987 on what standards should be used globally, the group recommended the original manometric method of Keeling. Since then, the CO_2 observing community has stuck to their own standard, which since 1995 has been maintained by NOAA.

The basis for expanding the ocean and atmospheric observation network was laid in the 1980s. In the atmosphere the coordination was taken up by the Global Atmospheric Watch programme from WMO, that built successfully on the initial US networks set-up by Keeling and NOAA. While major areas in the world are still under-sampled, an explosion of measurement sites has occurred more recently in Europe and the US (Figure 9.6). This is largely due to the increasing awareness of climate change and the need to get adequate observations of CO_2 on which to build a better understanding of the carbon cycle and eventually to monitor progress in the reduction of CO_2 emissions that is required by, for instance, the Paris agreement. Compared to the pristine sites that were the basis of the original network, many more land stations are now included. In Europe for instance, the network is run by a separate Research Infrastructure (Integrated Carbon Observing System, or ICOS) that has established strict protocols that enforce uniformity in both hardware, software, and data quality control. European countries can become members of ICOS and run a site according to the prescribed protocols. The data is collected and quality controlled at specified central data centres. ICOS runs

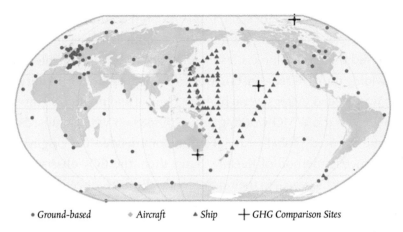

• *Ground-based*   • *Aircraft*   ▲ *Ship*   ✛ *GHG Comparison Sites*

Figure 9.6 Current global atmospheric network for measuring CO_2 in the last decade.
After WMO GAW, courtesy Oksana Tarasova (WMO)

such centres for atmospheric data, direct observations of the flux of CO_2 from ecosystems, ocean observations, radiocarbon, stable isotopes in CO_2, and other trace substances analysis on flask samples, and for calibration. To a large extent the build-up of these newer networks, that have also been established in China and Japan, was facilitated by a technological breakthrough: the development of cavity ring-down spectroscopy employing high accuracy infrared lasers. These instruments are capable of achieving the required stability and accuracy of WMO protocols at ±0.1 ppm, and in the long term may have higher accuracy and reliability than the traditional methods used for continuous monitoring. They also suffer less from water vapour contamination as the lasers used have a very high precision in wavelength, effectively only sampling the absorption by CO_2.

While Takahashi used 470 hours of measurements to arrive at his initial conclusions about global ocean uptake and releases, in

2014 Dorothee Bakker, an UK-based ocean scientist, published the latest collection of observations of surface pCO_2 concentrations. This Surface Ocean CO_2 Atlas (SOCAT) is now updated annually. By then the total number of observations had increased to a staggering 10.4 million, and in 2020 it had risen to 30.6 million observations! This growth in data collection resulted from the big international research programmes, such as the Joint Global Ocean Flux Study and the World Ocean Circulation Experiment, but also from regional funding initiatives. Data was collected from research vessels but also from 'ships of opportunity' that carry small, sophisticated measurement packages that collect data on their commercial sailings. A similar compilation of interior ocean data is given by the Global Ocean Data Analysis Project (GLODAP). Knowing where the anthropogenic carbon ends up in the deeper ocean is, of course, fundamental to understanding the longer-term behaviour of the ocean sink and its possible feedback on climate. The massive increase in data has made it possible to study not only the global annually averaged uptake and release of CO_2 by oceans, but also the seasonal variability and the differences between basins.

A final important development in observing CO_2 is the innovation in satellite sensor technology that have made it possible to detect atmospheric CO_2 mixing ratios directly from space. These measurements, again, make use of the absorption characteristics of CO_2 in the infrared and provide integrated values of CO_2 mixing ratio over the atmospheric depth (column). The accuracy is typically lower than the in situ ground-based system, but the big advantage is the amazing number of observations that can be retrieved. The first space-based mission specifically designed for measuring greenhouse gases was the Japanese Greenhouse gases Observing SATellite (GOSAT), also known by its Japanese nickname Ibuki (breath), that was launched in 2009. NASA's first dedicated mission, the Orbiting Carbon Observatory-1 (OCO-1), suffered a dramatic failure when the satellite failed to separate from the carrier rocket. The payload fairing that protects the payload on the launchpad and during the flight of the launch vehicle failed to separate. When the OCO satellite was finally separated from the launcher, it unfortunately still carried the fairing and because of the additional weight it then failed to reach the required orbit height. This resulted in a catastrophic atmospheric re-entry. The satellite valued at US$208 million broke up and/or burned up in the atmosphere. Surviving pieces were dispersed in the Pacific Ocean near Antarctica.

NASA quickly decided to build and launch a copy of the lost satellite, OCO-2, and this was successfully launched on 2 July 2014. Since then it has provided continuous data. A specifically designed sensor package, OCO-3, was launched on 3 May 2019 and placed on the International Space Station a few days later. This continues the measurements from OCO-2 but at a different orbit with a different viewing angle. Various other satellites are in the design stage and are expected to be launched over the next 5–10 years. These include satellites from Europe, and from China and Japan. With these new instruments, and importantly, the ground and ocean-based observations, a new era in observing the atmosphere's CO_2 content has arrived. Determining the emission reduction from burning less fossil fuel and the response of the ocean and land has become a realistic prospect.

CHAPTER 10

CLIMATE CHANGE, MODELS, AND THE ALLOWABLE CARBON BUDGET

W*e dive into the development of climate models: computer codes that model how the atmosphere works. The 2021 Nobel Prize winner Syukuro Manabe realized in 1967 that absorbed radiation also has to be emitted in the atmosphere and that at the top of the atmosphere, the incoming amount of energy has to be balanced by the outgoing amount. Manabe and his co-worker Wetherald then predicted that a doubling of CO_2 would lead to an increase in temperature of 2.3 °C. Manabe's model was originally just in the vertical (1-D). Soon siblings of numerical weather prediction models, that predicted the evolution of pressure, temperature, and humidity around the world, both horizontally and vertically (3-D), became available. This made it possible to implement Manabe's scheme in such a model: the first General Circulation Model was born. And the model produced some key insights that still stand today, such as polar amplification (the relative increase in temperature is greater near the poles with climate warming than in the tropics) and the acceleration of the hydrological cycle with climate warming. They predicted a warming of 2.9 °C for a doubling of the amount of CO_2 (600 ppm compared to 300 ppm).*

The results of these simulations slowly began to find their way into reports to policymakers, starting with those that were made in preparation for a 1972 UN meeting in Stockholm. The recommendations called for increased monitoring efforts. In 1979 Jules Charney put together the latest findings of what were by then several modelling groups around the world. He concluded that it was likely that the Earth would warm by 3 °C if the CO_2 concentration doubled. WMO had organized its first World Climate Conference and in the 1980s several other conferences and workshops followed, among them the 'Villach Conference' that was the first to call on governments for action to halt climate change. This advice from independent scientists was institutionalized in the Intergovernmental Panel on Climate Change that was established in 1988 and that has since, in 5–6 yearly cycles, provided advice to the governments of the world. Increasingly these reports were getting stronger in establishing the role of humans (read greenhouse gases such as CO_2) in changing the world's climate. In 2007 the IPCC was awarded the Nobel Prize for Peace together with Al Gore. Whether they were successful enough 'to contribute to a sharper focus on the processes and decisions that appear to be necessary to protect the world's future climate, and thereby to reduce the threat to the security of mankind' as the Nobel Prize Committee wrote, remains to be seen. Initially the uptake of the key messages, in any case, was rather slow and governments still do not always appreciate the implications of the reports.

Development of climate models continued and around 2000 several modelling groups started to incorporate a carbon cycle into their models. This was an important breakthrough as it allowed the natural carbon cycle to interact with emissions from fossil fuel. Since several of the natural carbon cycle processes, such as photosynthesis and respiration, are temperature dependent, it means that the natural carbon cycle can shift and further impact atmospheric concentrations. The main result was that even higher temper-atures could occur due to the interaction of the natural carbon cycle with climate. However, having these sophisticated models also allowed scientists to determine more precisely how much burned carbon would lead to a partic-ular temperature in the future. This has put a firm limit on how much fossil fuel humans can still burn (the remaining, or allowable carbon budget) and consequently how much we can emit, to remain below, say 1.5 °C or 2 °C. These numbers are small compared to what the world currently emits. Time is running out.

It was a stroke of luck that in the 1960s theoretical breakthroughs in meteorology and new developments in computing coincided with the launch of the first environmental satellites. In a speech to the United Nations on 25 September 1961, President John F. Kennedy keenly realized the opportunity this presented to make the world a more peaceful place: *'As we extend the rule of law on earth, so must we also extend it to man's new domain—outer space. All of us salute the brave cosmonauts of the Soviet Union. The new horizons of outer space must not be driven by the old bitter concepts of imperialism and sovereign claims. The cold reaches of the universe must not become the new arena of an even colder war. ... We shall propose further cooperative efforts between all nations in weather prediction and eventually in weather control. We shall propose, finally, a global system of communications satellites link-ing the whole world in telegraph and telephone, and radio and television. The day need not be far away when such a system will televise the proceedings of this body to every corner of the world for the benefit of peace.'*[1] The UN adopted a resolution that requested the World Meteorological Organization to study measures that would advance the state of

---

[1] Kennedy, J., 1961

atmospheric science and technology with the aim to improve weather forecasting and, and this is important in hindsight, 'to further the basic physical processes that affect climate'[2]. In response, WMO set up a task force headed by the directors of the US and USSR weather bureaus, Harry Wexler (who played an important role in establishing the CO_2 Mauna Loa record) and Victor Bugaev. This resolution formed the starting point of the World Weather Watch programme of WMO, which has in the long term significantly improved weather data exchange and improved the quality of weather forecasting. However, several scientists were also concerned that the operational drive to improve weather forecasting would not be possible without accompanying fundamental scientific research. Among these, Bert Bolin and Jules Charney pushed for a complementary research programme that would make use of the data provided by the new satellites. After consultation with other scientific bodies in the International Council of Scientific Unions, a new interunion Committee on Atmospheric Sciences was created in 1963 with Bolin as its first chairman. The CAS, with hindsight, was the first critical development in the establishment of a series of international research programmes in which East and West would find each other in collaborative efforts to improve the understanding of weather and climate, and ultimately the Earth System. CAS recommended the establishment of the Global Atmospheric Research Programme that started formally in 1967. This was later to evolve into the World Climate

---

[2] See Bolin, B., 2007, and United Nations, General Assembly, 1961. Interestingly the resolution not only called for the study of climate physics, but also called for research aimed at large-scale weather modification. Bolin's memoir provides a nice first-hand account of the early years of climate research and the history of the IPCC.

Research Programme.[3] Had he lived, Kennedy would have been pleased.

The year 1967 also saw the publication of a hugely important paper by the Japanese scientist Syukuro Manabe with his co-worker Richard Wetherald.[4] A survey among scientists found it to be the most influential paper in climate science—this paper came even before the Keeling paper that described the increase in CO_2 from the Mauna Loa measurements.[5] Manabe was awarded the Nobel Prize for physics in 2021, together with Klaus Hasselman, for their seminal contributions to climate modelling. Manabe realized that the issue of absorption of long-wave radiation by CO_2 was only part of the story of climate change. According to Kirchhoff's law (Chapter 2), when absorption increases, emission also has to increase. While Callendar was primarily concerned with the absorption part of CO_2 in the lower troposphere and concentrated on the surface energy budget, Manabe realized that the whole atmosphere would also have to adjust. This is because at the top of the atmosphere the incoming solar radiation has to be balanced by the outgoing long-wave radiation. It is a fundamental law in physics, known as the conservation of energy, that energy does not get lost. So, if the lower atmosphere warms up, it will emit more radiation upwards, and the upper atmosphere also warms up; the level of effective emission into space thus becomes located at higher altitude. It is this atmospheric response that is key to understanding the Greenhouse Gas effect and Manabe was the first to put some numbers on this response: *'So I was the first*

---

[3] Zilman, J., 2009
[4] Manabe, S. & Wetherald, R., 1967
[5] Carbon Brief, 2015

one to develop combined radiation and convection, and then clarify this effective source of emission issue, and also discuss what water vapor does to the outgoing radiation emissions. The water vapor will have an amplifying effect.'[6] He used a model that only described the atmosphere in one dimension, the vertical. But importantly, he included the effects of moisture, radiation, and convection (heat transport through movement of air parcels) in a single model. These models are known as radiative-convective models. Using this setup Manabe and Wetherald predicted that a doubling of CO_2 would lead to an increase in temperature of 2.3 K. What makes this paper so important is that it included the water vapour feedback by keeping the relative humidity constant rather than the absolute humidity. Importantly, they also predicted that a warming troposphere would lead to a stratospheric cooling. While the first issue may sound technical, it allowed a much more realistic response of the atmosphere because it incorporated the effects of temperature on the water-holding capacity of the atmosphere. The second provided a classic scientific prediction that was later borne out by observations of stratospheric temperature: the stratosphere indeed cools, while the troposphere warms due to increased concentration of CO_2.

Manabe's model was a one-dimensional representation of the most important physics of the atmosphere. The next phase was to develop a full three-dimensional representation of the atmosphere that included this physics. The first such model was formulated in 1955 by Norman Phillips at the Institute for Advanced Studies in Princeton, US. While his model had no moisture and clouds, his model was able to reproduce some of the main

[6] Manabe, S., 1989

circulation characteristics of the atmosphere, such as a jet stream, trade winds in the tropics and westerlies outside the tropics. He achieved this by splitting the atmosphere into two layers along the circumference of a cylinder, divided into 16 by 17 grid cells.[7] By running this model on one of the first computers at Princeton, the ENIAC (Electronic Numerical Integrator and Computer), he was able to generate realistic-looking weather patterns over a period of about 20 days. After that, the model essentially blew up on the computer.[8] Nevertheless, the experiment was, certainly with hindsight, hailed as the first proper Global Circulation Model experiment. In the years immediately afterwards, several other groups started to develop similar models, amongst them several in the US and the UK Meteorological Office. In the US at the Weather Bureau a new group was formed under the leadership of Joseph Smagorinsky. It was Smagorinsky who invited Manabe from the Tokyo Numerical Weather Prediction group to come and work with him on the development of the new GCM. The rest is history, with Manabe becoming one of the greats of GCM development and climate research.

The 1967 paper by Manabe and Wetherald was not aimed at calculating just the effect of CO_2; rather, it was meant as a further study of how to incorporate realistic models of radiative transfer into a general circulation model. But since one of their colleagues, Fritz Möller, had obtained rather surprising results, they decided to also use their model to calculate the CO_2-effect. Möller had shown that a doubling of CO_2 might lead to a 10 °C

---

[7] See for an excellent and extensive history of climate research and specifically the role of models, Edwards, P., 2010

[8] Phillips, N., 1956. See also Weart, S., 2003 for an extensive, although somewhat US centric, analysis of the development of global warming research.

change in global temperature. This was not only due to the way he incorporated humidity into his model, but his main problem was that he ignored the upper atmosphere heat balance. Something that Manabe and Wetherald had now corrected. The time was ripe for implementing Manabe's scheme in a GCM. Figure 10.1 shows the structure of the model.[9] For clarity, the arrows are kept as in the original, although some of the arrows in today's GCMs would be a double pointed rather than a single arrow. Also shown is where CO_2 impacts the climate: it affects only the radiative part of the code. The model was based on a set of earlier studies where, for instance, the impact of land surface hydrology in providing moisture to the atmosphere was incorporated. It was shown that

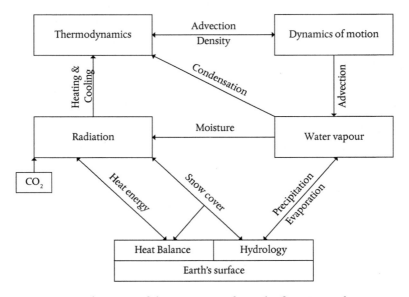

Figure 10.1 Schematic of the structure of one the first General Circulation Models showing the key elements.
Redrawn after Manabe, S., 1969

[9] Manabe, S. & Wetherald, R., 1967

the GCM reproduced the main features of the climate, such as a precipitation belt in the humid tropics, a dry belt over the world's deserts and important elements of heat transport from the Equator to the poles. The model was a very simple representation of the Earth's climate. The world was divided up almost evenly into land and ocean and the model did not include transport of heat by ocean currents. It did have nine layers in the vertical, which was a significant improvement over the earlier one- and two-layer models used by Philips and Charney. The model contained the effects of CO_2 and O_3 (ozone). By putting the effect of CO_2 in a three-dimensional context, Manabe and Wetherald were able to investigate the effects in a much more spatially representative way. This yielded immediately important new insights such as that near the poles the warming of the troposphere is larger than near the Equator: a phenomenon called polar amplification, something we appear to see playing out in full strength in recent years. They explained this by the effects of a reduced snow cover which led to more absorption (less reflection) of radiation and the fact that near the poles, the atmospheric boundary layer is relatively shallow, so that all the heat goes into a smaller layer than in the tropics, thus warming up the polar troposphere much more quickly. These days, the effect of decreasing sea ice is also thought to be a main contributor to polar warming. They also reproduced the two main features of their one-dimensional study: warming of the troposphere and cooling of the stratosphere. They obtained for a doubling of the CO_2 content (600 ppm) a globally averaged temperature increase of 2.9 °C, about half a degree more than their one-dimensional study, an increase attributed to the effect of snow cover, which was absent in their one-dimensional study. They also noticed an acceleration of the hydrological cycle.

Precipitation increased by 7% and also evaporation from the land surface increased by a similar amount. It is worth noting that their model used a fixed distribution of clouds, thereby ignoring one of the main feedbacks of the water cycle in climate. Later work would rectify this, but their main conclusions still stand today. Their study was a so-called equilibrium study. In such a study one runs two versions of the model until there is no change in the temperature. The standard model contains the (then) current CO_2 concentration (300 ppm) and the other double that concentration (600 ppm). By subtracting the results of these two runs, one obtains the effect of a doubling of CO_2 on the mean climate. It is a tribute to the efforts of Manabe and his group that the so-called equilibrium climate sensitivity (the increase of temperature for a given CO_2 increase) that they obtained is still well within the middle of the range of current estimates.

The late 60s and early 70s were a crucial period in getting environmental issues on the political agenda. With Rachel Carson having published in 1962, 'Silent Spring', the devastating account of the environmental impact of DDT, environmental issues began to get increasing attention. Through water and air pollution, pesticides in the environment, and acidification of surface waters, soil and forest were becoming increasingly impacted by the activities of humans. Under the title 'Restoring the quality of the environment', a US Presidential Advisory Committee was set up in 1965 under the leadership of John Tukey of Bell laboratories and including Roger Revelle to investigate a broad set of environmental problems.[10] A special subpanel under Revelle's leadership made recommendations on atmospheric carbon dioxide. The panel included the

---

[10] United States. President's Science Advisory Committee. Environmental Pollution Panel, 1965

now-familiar names: Charles Keeling, Joseph Smagorinsky, Wallace Broecker, and Harmon Craig. They stated, based on Keeling's observations that *The data show, clearly and conclusively, that from 1958 to 1963 the carbon dioxide content of the atmosphere increased by 1.36 percent. The increase from year to year was quite regular, close to the average annual value of 0.23 percent. By comparing the measured increase with the known quantity of carbon dioxide produced by fossil fuel combustion ... we see that almost exactly half of the fossil fuel CO_2 apparently remained in the atmosphere.*[11] They concluded, slightly rephrasing Revelle and Suess' famous 1957 quote, that *Through his worldwide industrial civilization, Man is unwittingly conducting a vast geophysical experiment. ... By the year 2000 the increase in atmospheric CO_2 will be close to 25%. This might be sufficient to produce measurable and perhaps marked changes in climate, and will almost certainly cause significant changes in the temperature and other properties of the stratosphere.*[12] It is interesting that they also, maybe for the first time, entertained the notion of geoengineering by suggesting that a possible countermeasure to the warming due to CO_2 might be to release *'particles'* or aerosols that would reflect the solar radiation. They further warned of impacts such as melting of the Antarctic ice cap, increased sea level, warming of the sea waters, and increased photosynthesis. Adequately predicting all these impacts has remained fully on the agenda of climate science and policy up to now.

Two other reports have been important in creating political awareness of climate and environment problems. They were both led by Caroll Wilson of the Massachusetts Institute of Technology and were set up to contribute to the planned 1972 UN conference

[11] United States. President's Science Advisory Committee. Environmental Pollution Panel, 1965
[12] United States. President's Science Advisory Committee. Environmental Pollution Panel, 1965

on the Human Environment. The SCEP report as it is generally known (report of the Study of Critical Environmental Problems) was the result of a one-month interdisciplinary study involving, again, a broad range of environmental problems[13] in ecology as well as climate. The report focused very much on the radiative effects that had recently been investigated by Manabe. They estimated that by the year 2000, the concentration of CO_2 could have approached 379 ppm (it was in reality 368 ppm) and resulted in a 0.5 °C increase in surface temperature (it was 0.4 °C in 2000, but 2000 was a slightly cold year, the years after were closer to 0.5 °C warming) with a stratospheric cooling of 0.5–1 °C. They also, and importantly, made a set of concise monitoring requirements specifying the accuracy and spatial and temporal resolution for measurements in both atmosphere and ocean, and on land.[14] They further called for research into carbon cycle processes in the main reservoirs of carbon, land and ocean. A subsequent report, the Study of Man's Impact on Climate (the SMIC), focused strictly on the climate issue only.

The actual UN conference of the environment took place in Stockholm, Sweden on 5–16 June 1972. A preparatory report was produced by Barbara Ward and René Dubos under the name of '*Only one Earth, the care and maintenance of a small planet*'; it was intended to provide a conceptual framework for the conference but found its way to the larger public as a commercial paperback.[15]

---

[13] MIT, 1970

[14] Having been strongly involved in monitoring myself, I am amazed by the level of detail of their foresight, and to some extent how little the key questions have changed, even though the monitoring techniques and computer power have changed dramatically. E.g. Ciais, P., et al., 2010.

[15] Ward, B. & Dubos, R., 1972.

The over 100 recommendations, some of which are very detailed and where climate, to some surprise, is only mentioned explicitly a few times, were grouped into three components: Earthwatch (environmental assessment) calling for increased research and monitoring, and information exchange; environmental management that included the precursor of what we now call ecosystem or environmental services, *'functions designed to facilitate comprehensive planning that takes into account the side effects of man's activities and thereby to protect and enhance the human environment for present and future generations,'*[16], and several supporting measures as the third component. Recommendation 79 for instance, called for '... *approximately 10 baseline stations be set up, with the consent of the States involved, in areas remote from all sources of pollution in order to monitor long-term global trends in atmospheric constituents and properties which may cause changes in meteorological properties, including climatic changes; (b) That a much larger network of not less than 100 stations be set up, with the consent of the States involved, for monitoring properties and constituents of the atmosphere on a regional basis and especially changes in the distribution and concentration of contaminants; (c) That these programmes be guided and coordinated by the World Meteorological Organization; (d) That the World Meteorological Organization, in co-operation with the International Council of Scientific Unions (ICSU), continue to carry out the Global Atmospheric Research Programme (GARP), and if necessary establish new programmes to understand better the general circulation of the atmosphere and the causes of climatic changes whether these causes are natural or the result of man's activities.'*[17] Global monitoring of properties and components of the atmosphere and GARP were now set in full swing and climate monitoring was put firmly on the agenda.

[16] United Nations, General Assembly, 1972.
[17] United Nations, General Assembly, 1972

With the advance of understanding the modelling had provided and with the observations at Mauna Loa providing clear evidence of a growth in atmospheric CO_2, a counterargument now started to appear, that seems to be the first of the climate deniers' attacks on the concept of CO_2-induced global warming. The issue centred around the impact of particle pollution in the atmosphere (aerosols). Ichtiaque Rasool and Stephen Schneider had published a calculation of the effects of these tiny particles on the surface temperature and found that these would act to cool, and in fact could well produce a global cooling of as much as 3.5 °C, enough to counter a CO_2-induced warming.[18] The paper contained several flaws, among them the absence of humidity feedbacks in their model, but stimulated discussion on whether the absence of large volcanic eruptions in the early half of the twentieth century could have contributed to the observed warming in the SMIC group. While the initial calculations of Rasool and Schneider proved erroneous, the precise role of aerosols in regulating climate is still one of the major uncertainties in the current climate models. The principal finding that (anthropogenic) aerosols have an important cooling effect still stands.

By 1975 several groups had developed a general circulation model. In the US, the National Center for Atmospheric Research had started to develop their own model and the Princeton Geophysical Fluid Dynamics Laboratory continued with the development of the original Phillips model. The University of California Los Angeles had also started the development of a GCM that was later used by the RAND corporation, and the Goddard Institute for Space Studies. In Europe the European Centre for Medium Range

---

[18] Rasool, S. & Schneider, S., 1971

Weather Forecasting, established in 1974 to improve weather forecasts up to 10 days ahead, built their own improved model from the GFDL model, and the UK Met Office had also started to develop their own model.[19] Later, in Germany, the newly established Max Planck Institute for Meteorology, under the leadership of Klaus Hasselmann, Manabe's co-winner of the 2021 Nobel Prize, took the atmospheric part of the ECWMF code to develop their own GCM. In one of the more recent model intercomparison projects, the Coupled Model Intercomparison Project Phase 5 (CMIP5), around 30 models from as many groups participated.

The Charney report, as it is known, is one of the first reports that set out an argued value for the likely greenhouse warming. Charney was a famous meteorologist working at MIT who had established, among others in the 1950s, the quasi-geostrophic approximation (an approximation of atmospheric flow that works at large scales) that formed part of the first generation of GCMs. He also became famous for his study of the importance of albedo (the reflection coefficient of sunlight) in establishing climate and drought in the Sahel. He was asked by the White House Office of Science and Technology to prepare a report on the climate issue. The group formed under his chairmanship also involved Bert Bolin and two oceanographers, Henry Stommel and Carl Wunsch. They analysed the results of six GCMs and concluded that the most probable warming for a doubling of CO_2 would be near 3 °C with a probability error of ±1.5 °C.[20] They highlighted the importance of the ocean as a 'thermal regulator' and suggested that this could delay noticeable warming of the atmosphere by several decades. They also emphasized that, although CO_2 was

[19] Edwards, P., 2010
[20] Charney, J., et al., 1979

well mixed in the atmosphere, the impacts of climate change could exhibit stark regional differences. They concluded that *'we tried but have been unable to identify any overlooked or underestimated physical process that could reduce the current estimated global warming due to a doubling of atmospheric CO_2 to negligible proportions or reverse them altogether.'*[21] Climate warming was going to happen, although the precise timing was still an issue, given the role of ocean heating.

In the same year, WMO had held its first World Climate Conference. It was organized in collaboration with the United Nations Educational, Scientific and Cultural Organization (UNESCO), the Food and Agriculture Organization of the United Nations (FAO), the World Health Organization (WHO), the United Nations Environment Programme (UNEP), the International Council of Scientific Unions (ICSU), and other scientific partners, as 'a world conference of experts on climate and mankind'.[22] The conference called for taking full advantage of increased knowledge of climate, taking steps to improve that knowledge, and foreseeing and preventing potential man-made changes in climate that might adversely affect the well-being of humanity. The Congress of WMO, the decision-making body of WMO's members, following the conference, established the World Climate Programme, that had as one of its components the successor to GARP, the World Climate Research Programme, chaired by Joseph Smagorinsky. The WCRP was a joint undertaking between ICSU and WMO, providing a link between the UN system, that was with WMO more tuned towards operational weather forecasting, and the more academic environment of the scientific community as organized in the non-governmental ICSU. If you are getting tired of

[21] Charney, J., et al., 1979
[22] Zilman, J., 2009

all the names, acronyms, and organizations that is quite under-
standable. However, this expansion of committees and groups is
also a sign that climate change was attracting the attention of an
increasing and fast-growing community, outside the traditional
weather services.

In the 1980s several other conferences and workshops were
organized to set out the challenge that increasing CO_2 meant for
climate and society. Following two smaller workshops in September 1980 and October 1982, a large conference, now generally
referred to as the 'Villach Conference' was held in the Austrian
city of Villach in Carinthia. It was organized by UNEP, WMO,
and ICSU. Villach is a medium-sized city beautifully located in
the Austrian Alps on a large lake. The conference was important,
maybe not so much for its new scientific findings—these were
in fact very much in line with the previous reports—but for the
fact that it called on governments to take action: '*Governments and
regional inter-governmental organizations should take into account the
results of this assessment (Villach 1985) in their policies on social and economic development, environmental programmes, and control of emissions
of radiatively active gases.*'[23] They also explicitly mentioned communication to a wider audience: '*Public information efforts should
be increased by international agencies and governments on the issues of
greenhouse gases, climate change and sea level, including wide distribution of the documents of this Conference (Villach 1985).*'[24] This call for
preventive action was something the World Climate Conference
had refrained from, as its prime recommendation was to further

[23] Co-ordinating Committee on the Ozone Layer, 1986
[24] International Conference on the Assessment of the Role of Carbon Dioxide and of
Other Greenhouse Gases in Climate Variations and Associated Impacts, et al., 1986. See
also Franz., W., 1997-1 & -2.

understanding through research. The conference statement of Villach summed it up concisely: 'While some warming of climate now appears inevitable due to past actions, the rate and degree of future warming could be profoundly affected by governmental policies on energy conservation, use of fossil fuels, and the emission of some greenhouse gases.'[25] Based on extensive preparations leading to the publication of a book,[26] the science on CO_2 and the implications were becoming increasingly clear. But there was another, probably more important issue at stake in Villach. Unlike previous reports, which tended to be rather nationally oriented, the Villach report appears to shed those restraints and calls for international action, something made possible through the fact that the participants were selected on the basis of their personal profile and not governmental background. Where previously the bounds of national policies provided constraints on calls for action, now the discussions among the participants did lead to these calls. The mechanism of having independent scientists providing assessment and advice would form to large extent the basis of the Intergovernmental Panel on Climate Change (IPCC), to be created a few years later in 1988.

The summer of 1988 was very hot in the US and several heatwaves and droughts plagued the country. From a climate policy perspective, the summer was important because NASA Goddard Institute for Space Sciences director James Hansen testified to a congressional hearing. Weart[27] suggests that the exact timing of

---

[25] Co-ordinating Committee on the Ozone Layer, 1986
[26] Bolin, B., et al., 1986. SCOPE 29: The Greenhouse Effect, Climate Change and Ecosystems. SCOPE is the Scientific Committee on Problems of the Environment of ICSU that issued a series of ground-breaking reports, including two reports on the Global Carbon Cycle (SCOPE 13 in 1979 and SCOPE 62 in 2004).
[27] Weart, S., 2003

the hearing during the hot summer was deliberate to emphasize the importance of climate change. Hansen stated that 'with 99% confidence' a long-term warming trend was underway and likely caused by increases in greenhouse gases. The New York Times opened the next day with the headline 'Global warming has begun expert tells Senate' and showed a graph of increasing temperatures from 1880 till 1988.[28] Manabe confirmed that global warming would exacerbate the droughts in the future. Other members of the panel, such as George Woodwell from the Woods Hole Research Center called for immediate reduction in the use of fossil fuels and for stops on deforestation. The heat was on, and research showed that a year later the amount of people who had heard about the greenhouse effect doubled from almost 38% to 79%.[29] While Hansen's testimony has achieved almost mythical status in climate history by now, it is important to note that not everybody that was involved in climate research at the time was on the same page. Take Bert Bolin in his memoir: '*An intense debate amongst scientists followed and most of them disagreed strongly with Hansen's statement. The data showing the global increase of temperature had not been scrutinised well enough and there was insufficient evidence that extreme events had become more common. This was to me a clear warning of how chaotic a debate between scientists and the public might become, if a much more stringent approach to the assessment of available knowledge was not instituted.*'[30] About a year earlier both UNEP and WMO had agreed to set up an intergovernmental assessment panel on climate change. Such a panel was supposed to bring some sort of harmony and consensus into the various studies that

---

[28] New York Times, 24 June 1988.
[29] Weart, S., 2003
[30] Bolin, B., 2007

were now appearing. The panel was called the Intergovernmental Panel for Climate Change, commonly known as the IPCC. In the UN Assembly in December 1988, resolution 43/53, packed in between resolutions about, for instance, the Middle East situation and apartheid in South Africa, the UN called for the protection of global climate for present and future generations of mankind. It defined the task of the IPCC as '*to prepare a comprehensive review and recommendations with respect to the state of knowledge of the science of climate change; the social and economic impact of climate change, and potential response strategies and elements for inclusion in a possible future international convention on climate.*'[31] The three items in the original resolution were to be elaborated in three different working groups: science, impact, and mitigation. Basically, this structure has remained the core of the IPCC until now.

Bert Bolin (Figure 10.2) was asked to chair the IPCC and Sir John Houghton, director of the UK Meteorological Office was appointed chair of Working Group I, the assessment of the scientific basis. The first report was due to be presented at the UN General Assembly in 1990. This gave a short time frame of only 18 months for the group to prepare a first report. Around the same time, the political world also started to become increasingly interested in the issue of climate change. In the UK, for instance, Margaret Thatcher realized the importance of climate change and the threat it posed to our natural habitat in a speech to the Royal Society. In the coastal town of Noordwijk in the Netherlands, an international conference was held which drew many world leaders. Whether accumulating science was behind these developments making politicians more aware of environmental issues,

---

[31] United Nations, General Assembly, 1989

Figure 10.2 Bert Bolin, the first chairman of the IPCC and one of the first to warn humankind of dangerous climate change.
Photo credit: The Royal Swedish Academy of Science

or whether there was a short-term process with politicians getting benefits by changing their attitude is still the subject of debate among historians.[32]

The establishment of the IPCC marked a definite shift away from the local, meteorological aspects of weather and climate as a statistical description, to a more global climate understanding with all the domains of the Earth System playing an important part.[33] In this view the risks associated with an increase in CO_2

---

[32] Agar, J., 2015 gives considerable detail on the behind-the-scenes development and the establishment of the UK centre for climate research, the Hadley Centre. See also Mahony, M. & Hulme, M., 2016.

[33] Miller, C., 2004

were global, and could only be understood, and managed, at the global level.

In their first report of 1990, the IPCC noted with certainty that *There is a natural greenhouse effect which already keeps the Earth warmer than it would otherwise be. Emissions resulting from human activities are substantially increasing the atmospheric concentrations of the greenhouse gases: carbon dioxide, methane, chlorofluorocarbons (CFCs) and nitrous oxide. These increases will enhance the greenhouse effect, resulting on average in an additional warming of the Earth's surface. The main greenhouse gas, water vapour, will increase in response to global warming and further enhance it.*[34] They stated that *'Carbon dioxide has been responsible for over half of the enhanced greenhouse effect in the past, and is likely to remain so in the future. Atmospheric concentrations of the long-lived gases (carbon dioxide, nitrous oxide, and the CFCs) adjust only slowly to changes of emissions. Continued emissions of these gases at present rates would commit us to increased concentrations for centuries ahead. The longer emissions continue to increase at present day rates, the greater reductions would have to be for concentrations to stabilize at a given level.*'[35] Importantly, they also noted the interaction of the natural carbon cycle with the anthropogenic emissions: *The human-caused emissions of carbon dioxide are much smaller than the natural exchange rates of carbon dioxide between the atmosphere and the oceans, and between the atmosphere and the terrestrial system. The natural exchange rates were, however, in close balance before human-induced emissions began; the steady anthropogenic emissions into the atmosphere represent a significant disturbance of the natural carbon cycle.*'[36] In a nutshell this is the core content of the IPCC reports up to the 6th assessment report that came out in 2021.

---

[34] IPCC, 1990
[35] IPCC, 1990
[36] IPCC, 1990

Also, the IPCC-speak of 'with certainty' and 'with confidence' would become an increasing part of the rhetoric of the IPCC. In its work on predicting the future the IPCC relied heavily on the newly developed Global Circulation Models. Noting the many uncertainties due primarily to lack of knowledge, they concluded the 1990 report with the judgement that: '*The size of the warming over the last century is broadly consistent with the prediction by climate models, but is also of the same magnitude as natural climate variability.*' In other words, in 1990 they could not rule out the possibility that natural variability, and not an increase in carbon dioxide, was the main cause of the observed 0.3–0.6 °C increase in global temperature over the previous century. Later reports would come back on this issue.

How could the IPCC become so successful, that in 2007 it was even awarded the Nobel Prize for Peace, together with Al Gore? From the press release of the announcement of the price: '*By awarding the Nobel Peace Prize for 2007 to the IPCC and Al Gore, the Norwegian Nobel Committee is seeking to contribute to a sharper focus on the processes and decisions that appear to be necessary to protect the world's future climate, and thereby to reduce the threat to the security of mankind.*' The Nobel Committee linked climate change to security and hence peace in a way reminiscent of John F. Kennedy earlier linking the development of satellites and weather. Analysis of the modus operandi of the IPCC also reveals several factors that contributed to its success.[37] The first of these is the structure of the organization of the IPCC. A Bureau, located at WMO in Geneva, and technical support groups for the working groups organize the logistics of bringing all the scientists together and help preparing the reviews

---

[37] Miller, C., 2004

and reports. In the operation, transparency and democracy feature widely to boost the credibility and authority. Secondly, there is a marked distinction between the scientists who participate in the process and the country delegates who approve the report. While the required approval of the participating governments, particularly in the writing of the Summary for Policymakers that accompanies each report, clearly reflects an increasing level of government participation compared to say, the book from the Villach conference,[38] it also provides the IPCC with a legitimacy that would have been hard to obtain if it was just a pure scientific undertaking. That is not to say the IPCC is without criticism. One of the more well-known gaffes, which turned out to be a simple typo, is the statement in the Fourth Assessment report on impacts, which says that Himalayan glaciers are likely to melt completely by 2035.[39] Another similar, somewhat careless mistake concerns the statement in the same report that the Netherlands has 55% of its country below sea level, whereas it was meant to say that 55% is prone to flooding with 29% lying under sea level. But also, its sea level projection itself came under scrutiny; here by the accusation that they were underestimating the potential contribution due to fast melting ice sheets in Greenland. While the first two are likely human errors, something that will nearly always happen with such a large undertaking, the sea level issue was more serious in that it seemed like a deliberate choice not to include the fast-melting processes. Others, more difficult to handle, were that the then chairman, Pachauri, had ties with companies that stood to benefit from IPCC recommended policies and was later accused of sexual harassment. But also, the selection of lead scientists that

[38] Bolin, B., et al., 1986
[39] Cruz, R., et al., 2007

head up the writing by governments remains a rather opaque process. On the other hand, it must also be said that virtually all of these lead authors are excellent scientists, well respected by their broader communities, even though some might boost their career by becoming a lead author. All these criticisms led the UN through UNEP, one of the founding organizations of IPCC, to ask the InterAcademy Council in 2010 to conduct a review of its procedures. The need for this review was probably made more urgent by an active public discussion on climate generated by the publication of a set of hacked emails by UK climate scientists. The controversy, also known as ClimateGate, generated a lot of attention and led to questions and investigations in parliament. In the end the involved scientists were cleared and exonerated.[40] The IAC after its review concluded that overall, things were going well, but could also be significantly improved: *The overall structure of the IPCC assessment process appears to be sound, although significant improvements are both possible and necessary for the fifth assessment and beyond. Key improvements include enhancing the transparency of the process for selecting Bureau members, authors, and reviewers; strengthening procedures for the use of the so-called "gray literature"; strengthening the oversight and independence of the review process; and streamlining the report revision process and approval of the Summary for Policymakers.*[41] They also made a recommendation that all working groups in the future should use a common quantitative uncertainty scale and made several recommendations to improve communication and management. IPCC responded in 2011 to the recommendations by introducing several new protocols and made several changes to the structure of its management.

---

[40] Adam, D., 2010
[41] Committee to Review the Intergovernmental Panel on Climate Change, 2011

The question now, of course, is: what have the series of IPCC assessments given us in terms of improved understanding of the role of carbon dioxide in climate? The Summary for Policymakers is the part of the report that is negotiated by the delegates as an adequate final summary of the scientific assessment; it is often concluded in a night-long session, where the final text is agreed sentence by sentence, word by word. The second assessment report made the famous statement about the shifting balance of evidence: '*Our ability to quantify the human influence on global climate is currently limited because the expected signal is still emerging from the noise of natural variability, and because there are uncertainties in key factors. These include the magnitude and patterns of long-term natural variability and the time-evolving pattern of forcing by, and response to, changes in concentrations of greenhouse gases and aerosols, and land surface changes. Nevertheless, the balance of evidence suggests that there is a discernible human influence on global climate.*'[42] In the 1990 report the scientists had maintained that the increase in global temperature could also fit within natural climate variability. Five years later, the IPCC scientists thought that the balance of the evidence suggested that man-made climate change was becoming a fact. Again, five years later, the Third Assessment report concluded that with '...*a longer and more closely scrutinised temperature record and new model estimates of variability. The warming over the past 100 years is very unlikely to be due to internal variability alone, as estimated by current models.*'[43] New attribution techniques, inclusion of aerosols in the models (remember the earlier discussion based on the Rasool and Schneider paper),

---

[42] IPCC, 1995. All other reports of the IPCC can be downloaded from their website https://www.ipcc.ch/reports/.
[43] IPCC, 1995

extensive model simulations with natural forcing (solar variability, volcanic eruptions) further supported the view that, *'In the light of new evidence and taking into account the remaining uncertainties, most of the observed warming over the last 50 years is likely to have been due to the increase in greenhouse gas concentrations.'*[44] The word 'likely' meant a 66–90% chance in the IPCC judgement criteria ('virtually certain' meaning a 99% chance that the statement is true). In the Fourth Assessment report from 2007, they left no doubt what was causing the increase in greenhouse gases: *'Global atmospheric concentrations of carbon dioxide, methane and nitrous oxide have increased markedly as a result of human activities since 1750 and now far exceed pre-industrial values determined from ice cores spanning many thousands of years. The global increases in carbon dioxide concentration are due primarily to fossil fuel use and land use change, while those of methane and nitrous oxide are primarily due to agriculture.'* There was also no doubt about the warming anymore: *'Warming of the climate system is unequivocal, as is now evident from observations of increases in global average air and ocean temperatures, widespread melting of snow and ice, and rising global average sea level.'*[45] In about 15 years we moved from 'hard to distinguish' to 'unequivocal'. In 2013 the report was dedicated to Bert Bolin, the first chairman of the IPCC, who had died that year in the last days of December. They stated again that *'Warming of the climate system is unequivocal, and since the 1950s, many of the observed changes are unprecedented over decades to millennia.... It is virtually certain that globally the troposphere has warmed since the mid-20th century.'* There was, further, little doubt that CO_2 increases provided the bulk of the changes in radiative forcing that warmed the planet: *'Human influence on the climate system is clear. This is evident from the increasing*

[44] IPCC, 1995
[45] IPCC, 2007

greenhouse gas concentrations in the atmosphere, positive radiative forcing, observed warming, and understanding of the climate system.'[46]

Around the 6th Conference of the Parties in the Hague a publication in *Nature* came out that showed that warming would be higher in a model that included an active carbon cycle. The paper contributed significantly to a development of Earth System Models in a direction where a much more realistic carbon cycle interacted with the emissions from fossil fuel and land-use change. In the paper[47] Peter Cox stated that '... *under a 'business as usual' scenario, the terrestrial biosphere acts as an overall carbon sink until about 2050, but turns into a source thereafter. By 2100, the ocean uptake rate of 5 Gt C yr^{-1} is balanced by the terrestrial carbon source, and atmospheric CO$_2$ concentrations are 250 p.p.m.v. higher in our fully coupled simulation than in uncoupled carbon models, resulting in a global-mean warming of 5.5 K, as compared to 4 K without the carbon-cycle feedback.'* It is important to understand what this means to appreciate the subsequent attention the paper got in the press. It effectively says that the natural carbon cycle can act to amplify human-induced warming. The region where this was most visible was the Amazon Forest where dieback of the simulated forest was triggered by the lack of rainfall and high temperatures. A tropical forest needs about 1800 mm of rainfall per year and when over several years the annual rainfall is less than that, the forest will shift towards a savannah. Later research has modified the conclusion of the paper substantially, as the original Hadley Centre model that was used had a so-called dry bias over Amazonia (it produced too little rainfall there). This bias probably contributed to the strong dieback that the vegetation model showed.

---

[46] IPCC, 2013
[47] Cox, P., et al., 2000

However, after this publication, and a simultaneous one from a French group led by Pierre Friedlingstein, the inclusion of a carbon cycle became of critical importance for all climate models. To show that importance, take a look back at Figure 10.1 where CO_2 only impacts the radiation part of the code and nothing else. The newer set of models used in the IPCC assessments later than 2013 now all include a carbon cycle, where CO_2 interacts with the vegetation (likely providing enhanced growth and later decay) and the ocean (providing enhanced acidity through the uptake of CO_2, up to a point…) as explained in previous chapters in this book. Models are ideal tools to study these processes as one can conveniently switch off particular feedbacks to investigate their strength compared to a model run where the feedback is kept intact. This allows us to investigate the response of the carbon cycle to an increase in the CO_2 concentration (the so-called carbon-concentration feedback) and the response to an increase in temperature alone (the so called carbon-climate feedback). For instance, in what is called the biogeochemically coupled simulation, biogeochemical processes such as photosynthesis, vegetation growth and change, and ocean uptake respond to increasing atmospheric CO_2, while the radiative transfer calculations in the atmosphere continue to use a CO_2 concentration that is kept at a pre-industrial value around 280 ppm. In contrast, in a radiatively coupled simulation, one can switch off the response of the biosphere to CO_2 by keeping it at a constant level for biogeochemical processes, whereas in the atmosphere CO_2 is still allowed to increase. Putting it simply, the latter is effectively how Manabe and his group started their first climate sensitivity experiments. Using eight models the mean value of the carbon-concentration feedback was found to be 0.97 ± 0.4 Pg C ppm^{-1} over land; over the ocean it was

0.79 ± 0.07 Pg C ppm^{-1}.[48] Note that since pre-industrial time we have added about 140 ppm CO_2 to the atmosphere, so to get an historical estimate these values should be multiplied by the increase in concentration (140 ppm). This makes the numbers comparable to those of the carbon-climate feedback, which are, noting that global temperatures have increased by 1.1 °C, expressed as an amount of carbon per degree temperature, -45 ± 50.6 Pg C °C^{-1} over land and -17.2 ± 5 Pg C °C^{-1} over sea, respectively. These carbon-climate effects are negative, similar to the dieback story of the Amazon, and significant. Overall, the feedback of the natural carbon cycle on human emissions is thus important, certainly in the longer term, as we are still increasing the atmospheric CO_2 concentration by 2.5 ppm yr^{-1}.

The carbon-concentration and carbon-climate feedbacks have opposing signs. Increasing CO_2 stimulates vegetation growth, in principle, but increasing temperature then can lead to increasing carbon losses from vegetation as (soil) respiration effectively outcompetes photosynthesis and vegetation such as that in the tropics may die because of droughts. However, for the next 50–70 years or so the storage of carbon on land is expected to increase. In the radiatively coupled simulations with no biogeochemical response, increasing temperature leads to a loss of carbon over the land (negative flux). If the biosphere is allowed to respond it takes up carbon and a positive carbon flux is obtained. In the fully coupled run, where both the radiation and the biogeochemical components are switched on, that positive flux is reduced by roughly the amount that the radiative coupled run predicts. The large uncertainty ranges in the numbers quoted above indicate

[48] Arora, V., et al., 2020

that the precise response depends very much on the model, this being a combination of an atmosphere, ocean, and land model.

However, there is a probably more important point to note. The biogeochemically coupled flux in the models appears to remain stable for about 50–70 years and then starts to bend downwards. This behaviour also holds for the fully coupled runs. What this means is that the amount of carbon the land biosphere sucks up is declining: the sink is losing strength. The models, by the way, simulate something similar happening in the oceans. This is bad news, because, as we have seen in the previous chapter, only about 45% of the emitted CO_2 from fossil fuel combustion ends up in the atmosphere. This is a consequence of the uptake of CO_2 by the ocean and land. If they stop doing that, the atmospheric CO_2 would increase further at an even higher rate. Predicting the timing and magnitude of this decline of uptake strength is, of course, an active area of research.

The introduction of the carbon cycle into Earth System Models has other implications that have also fed directly into key elements of the Paris agreement which we discuss in the next chapter. This brings us to something that is called the Transient Climate Response or the politically more important Transient Climate Response to cumulative Emissions (TCRE). The Transient Climate Response is different from what we earlier called equilibrium climate sensitivity, the ratio of the increase of temperature over a doubling of CO_2. The difference is in the word transient, implying that there is a gradual increase imposed of about 1% CO_2 increase per year in the model. In the equilibrium sensitivity runs, two runs were performed, one with a pre-industrial concentration of CO_2 and one with double that value. The surprise appears when we start summing up emissions and plot these cumulative emissions against the temperature from the transient simulations. It then turns out that there is a very nearly linear relation between these summed emissions and the maximum temperature, something that holds over a wide range of models and, in fact, observations. The reason for this behaviour appears to be a subtle balancing act going on in the models between the radiative forcing from higher atmospheric concentrations (this increases with the square root of the concentration, so a fourfold increase yields only a double increase in radiative forcing) and the expected decline in sink strength of the ocean and land that we just discussed.

Figure 10.3 shows how this works. On the temperature axis one can determine the maximum temperature allowed; on the x-axis one can then read off how much emitted carbon this would allow. This is the cumulative amount, so from the start of fossil fuel emissions is counted. Since there is uncertainty attached to the outcomes of the different models, the allowable carbon budget also has uncertainty. This is obtained, not by reading off at the solid line, but at one of the broken lines, that give a, say, 66% uncertainty range. The latest IPCC estimates in a special report on the implications of the 1.5 °C temperature[49] threshold that was agreed in the Paris Agreement finds that the total amount of carbon that can be put into the atmosphere since the industrial revolution is 604 Gton C (±88). Since we have already put in a substantial amount, the remainder is what should really concern us. This is of course much smaller: 159 or 211 Gton for a 50% probability—if one wants a higher probability then the numbers become smaller: 115 and 156 Gton C. We are currently putting in about 10 Gton C per year into

[49] IPCC, 2018

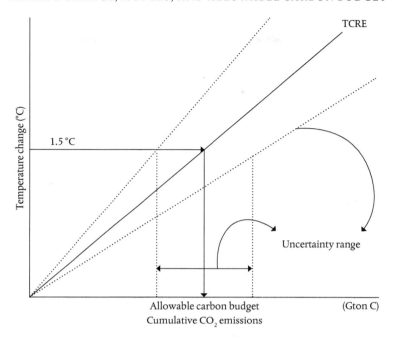

Figure 10.3 The concept of the Transient Climate Response to Emissions.

the atmosphere, and this is still increasing. Because other gases such as methane and nitrous oxide also act as greenhouse gases, the Transient Climate Response to cumulative Emissions' remaining carbon budget estimates, when based on CO_2 only, are likely to be underestimates. Importantly, other feedbacks from the Earth System, such as increased thawing of permafrost and its associated release of carbon could affect the carbon budget and make it smaller. How to handle this carbon budget and how to reduce the emissions, that if they continue at current rates would make holding to the 1.5 °C threshold virtually impossible, is the subject of the next two chapters.

CHAPTER 11

THE WORLD COMES SLOWLY INTO ACTION

*T*wenty years after the Stockholm conference, the world's leaders met again, now in 1992 in Rio de Janeiro, Brazil. They agreed on the establishment of the UN Framework Convention on Climate Change (UNFCCC). This convention was aimed to establish rules and regulations to combat climate change, based on the premise of 'common but differentiated responsibilities'. This held that countries that had emitted most historically, i.e., the USA, Europe, held a special responsibility to reduce most, while other countries on a path to development would be helped by those countries to achieve development without increased emissions. The convention went into force in 1994. The goal was to stabilize climate; at what level and how needed to be negotiated at annual Conferences of the Parties (the famous COP meetings, such as COP 26 in 2021 in Glasgow). In 1997 the Kyoto protocol was negotiated in which developed countries agreed to reduce their emissions compared to 1990. The USA, one of the largest emitters, never ratified the protocol, as the Senate with a 95–0 majority had already put a death sentence on any potential outcome of Kyoto that would put the blame one-sidedly on the developed world, or the US. Still, the rest of the world tried to make it work, with promises to reduce by 8% (European Union) compared to 1990 emissions by 2008–2012. With the arrival of George Bush as president, the US *withdrew altogether from the Kyoto protocol, likely influenced by powerful and insidious lobby campaigns by large oil firms like Exxon.*

The Kyoto protocol ended in 2012 and thus a new agreement needed to be negotiated that would start at that time with goals stretching towards 2020. The negotiations for this started in 2009 in Copenhagen at COP 15. While the expectations were high, partly because of increased public aware-ness of the climate issue, the meeting ended in a disappointment with the so-called Copenhagen declaration, brokered by the US and China, that con-tained nothing much that was legally binding. It did mention a 2 °C warming limit the world should strive for.

In 2015, the French organizers of COP 21 prepared the meeting well in advance and managed to secure a legally binding agreement, the Paris Agree-ment. This set out to limit the global temperature rise to no more than 1.5 °C and certainly below 2 °C. The Paris Agreement is a pledge and review agree-ment where countries have to submit their plans for emissions reductions and the total of those is then assessed in an open and transparent way as to whether they are bold enough to keep the temperature below the agreed thresholds. If not, countries need to show more ambition in their reduction plans in the next cycle. The Paris Agreement came into force in 2016, with a first assessment period, known as the Global Stock Take, to take place in 2023. In 2017 US President Trump withdrew from the agreement as soon as he was appointed, but four years later President Biden joined again, showing the fragile nature of the agreement.

In Glasgow, after two years of COVID restrictions the world leaders met again at COP 26 in 2021 to discuss how to implement the Paris Agreement and how to ensure transfer of funds to the developed world to help adapta-tion to climate change. While pledges were made to reduce emissions from deforestation and from methane, the big item remained how to deal with carbon dioxide. The discussion in the end was about whether to 'phase out' or 'phase down' fossil fuel burning. It was phasing down, a disappointment to many of the negotiators and certainly to the general public and activists that had become ever more vocal in their frustration at the slow pace at which the world was dealing with the climate crisis.

The year 2021 saw climate change imprinting its image firmly on the Earth with devastating heatwaves and fires in Canada, deadly floods in central Europe, and a plethora of other extreme events around the world. It was also the year in which the COVID-19 virus caused millions of deaths and large-scale lockdowns were imposed. Amidst this turmoil, countries were trying to agree for the 26th time in a Conference of the Parties (COP) on measures to keep the Earth's temperature rise manageable. COP26 took place in November in Glasgow in Scotland.

The global governmental response to climate change goes back to 1992 with the 'mother of all environmental congresses', held in Rio de Janeiro, Brazil. Twenty years after the Stockholm meeting (Chapter 10) that put climate change on the scientific agenda but not yet in direct view of policymakers, the United Nations Conference on Environment and Development welcomed over more than 100 heads of government. Led by Maurice Strong, who had also chaired the Stockholm meeting, the meeting was heavily influenced by a report of the Brundtland commission 'Our Common Future' published five years earlier. A commission, led by the Norwegian prime minister Gro Harlem Brundtland, had developed new notions on how to deal with an increasing world population and its environmentally damaging resource use. The Brundtland report was influential because it defined a solution to the two key problems in the world: one, that the richer part of the world was producing and consuming in a non-sustainable way, with damaging environmental impacts for future generations; and two, that the other part of the world lived in poverty without access to that wealth and resources. The solution the commission came up with was 'sustainable development', which it defined as *'development that meets the needs of the present without compromising the ability of future generations to meet their own needs.'*[1] While recognizing the sometimes deplorable state of the global environment, the commission believed *'that people can build a future that is more prosperous, more just, and more secure. Our report, Our Common Future, is not a prediction of ever-increasing environmental decay, poverty, and hardship in an ever more polluted world among ever decreasing resources. We see instead the possibility for a new era of economic growth, one that must be based on policies that sustain and expand the environmental resource base. And we believe such growth to be absolutely essential to relieve the great poverty that is deepening in much of the developing world.'*[2] They also noted, however, that the *'hope for the future is conditional on decisive political action now to begin managing environmental resources to ensure both sustainable human progress and human survival.'* Some of their optimism was derived from advances in technology: *'Fortunately, ... we can move information and goods faster around the globe than ever before; we can produce more food and more goods with less investment of resources; our technology and science gives us at least the potential to look deeper into and better understand natural systems. From space, we can see*

[1] World Commission on Environment and Development, 1987
[2] World Commission on Environment and Development, 1987

and study the Earth as an organism whose health depends on the health of all its parts. We have the power to reconcile human affairs with natural laws and to thrive in the process.'[3] Deemed by some as a success, others, such as the New Zealand Prime Minister of the time, Greg Palmer,[4] considered it a failure. Nevertheless, the 1992 meeting was important because nations were invited to sign up to three international conventions, of which the one of special interest to us is the UN Framework Convention on Climate Change, perhaps better known as *'UNF triple C'*. The other two were the Convention on Biodiversity (CBD) and the UN Convention to Combat Desertification (UNCCD).

UNFCCC is a convention, and this has implications for the way it operates. Within a convention protocol, individual countries first agree to collaborate; they later specify exactly how.[5] The choice for a convention was based on the success of the reduction of aerosol propellants that had been destroying the ozone layer. In 1985 the convention for the protection of the ozone layer was established in Vienna, which led in 1987 to the establishment of the Montreal Protocol that successfully worked to halt and reverse the further reduction of the ozone layer. In fact, the Montreal Protocol is often hailed as the prime singular example of an environmental agreement that did fulfil its promise. We will see that the path to CO_2 reduction was, and is, far more complicated. From 1991 to mid-1992, a precursor to the UNFCCC was formed that negotiated the text for the convention. A timeline of the most important conferences discussed in this chapter is given in Figure 11.1. This Intergovernmental Committee for a

[3] World Commission on Environment and Development, 1987
[4] Palmer, G., 1992
[5] Kuyper, J., Schroder, H., & Linnér, B., 2018

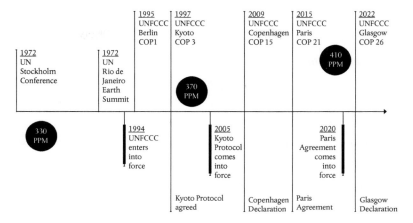

Figure 11.1 Timeline of the key COP meetings discussed in this chapter. Also shown is the relentlessly increasing concentration of CO_2 in the atmosphere.

Framework Convention on Climate Change met five times and was faced with the key problems of climate change negotiations: who was responsible for the historic emissions, which countries were the most vulnerable, and how could this complex difference in socioeconomic welfare be considered in a mutually satisfying way? The answer was the concept of 'common but differentiated responsibilities'.[6] This held that there was a common problem, the warming induced by greenhouse gases, but that countries who had emitted historically the most should bear the largest burden of reducing emissions, while countries in a lesser state of development would for now be exempt from these reductions. Equity between countries was thus at the heart of the convention, and from the start played a critical role in the negotiations.

When finally, the convention was opened up for signature in Rio de Janeiro, 135 countries signed up to it. What they signed up

---

[6] van der Gaast, W., 2017

to was a set of rules, consensus-based, that aimed: *'to achieve, in accordance with the relevant provisions of the Convention, stabilization of greenhouse gas concentrations in the atmosphere at a level that would prevent dangerous anthropogenic interference with the climate system. Such a level should be achieved within a time frame sufficient to allow ecosystems to adapt naturally to climate change, to ensure that food production is not threatened, and to enable economic development to proceed in a sustainable manner.'*[7] The overall objective was thus defined: make sure that the concentration of greenhouse gases did not reach levels where one could speak of dangerous implications for, or interference with, our climate system. What precisely the level of stabilization was, or when one could speak of dangerous interference, was still unclear. By June 1993 the Convention had received 166 signatures but could only enter into force when more than 50 parties had ratified it. This happened on 21 March 1994, ninety days after the 50th country's ratification had been received.

The 1992 Convention did not specify legally binding targets for reduction, let alone a compliance regime that could punish countries if targets were not met. It would take several more years until this was negotiated as the Kyoto protocol. Before that, the first Conference of the Parties (COP) was held in 1995 in Berlin, the new capital of the recently reunited Germany. Its main task was to review the commitment made by countries in preparing inventories of their emissions and policies to reduce emissions up to 2000. In these commitments the Convention differentiated between the developed countries (grouped as Annex-I) and the non-developed countries (grouped as non-Annex-I). Annex I countries included, amongst others, the US, Canada, the EU,

[7] United Nations, 1992

while non-Annex-I countries made up the rest of the signatories. In Berlin it appeared that, out of the Annex-I countries, nine countries expected an increase in CO_2 emissions up to 2000 compared to the agreed baseline of 1990, and only six countries would expect to see a reduction or stabilization. No surprise, then, that the conclusion was that the commitments were inadequate to realize the goal of the convention. This is something that happens now at almost every COP meeting: the term for this was coined during the Paris negotiations as 'the emissions gap'. This is the difference or gap between how much countries should reduce their emissions to achieve the goal of stabilizing climate and the commitments made on what individual countries will actually do.[8] Negotiations in Berlin were complicated, with several countries organized in large blocks with common goals and aims within the group, but opposed to other groups. In the end a text emerged, the Berlin Mandate,[4] which called for the development of specific measures and policies and the setting of limits and reduction targets for well-defined points in time, such as 2005, 2010, 2020. Importantly, this was an agreement to negotiate such targets, not an agreement on the target itself. They also agreed that these reduction targets would only apply to Annex-I countries, in line with the principle of common but differentiated responsibilities.

To take the negotiations further an ad hoc group was set up on the Berlin Mandate to provide the platform for the negotiations that led to the Kyoto Protocol at COP 3.[9] A flurry of diplomatic activity took place in the months just before the Kyoto meeting.

---

[8] Emissions Gap reports are prepared by UNEP, see United Nations Environment Programme, 2020
[9] Grubb, M., et al., 1999

Key issues on the agenda were still the establishment of country commitments, how these were differentiated, and what flexibility could be achieved in reaching these commitments: for instance, could these be affected outside the country's own territorial area? Discussions also included the issue of which greenhouse gases to include, not only CO_2, but also other gases, particularly methane and nitrous oxide, that were known to generate greenhouse warming. The protocol agreed on what became known as the basket of greenhouse gases: carbon dioxide, methane, nitrous oxide, hydrofluoride, perfluorocarbons, and sulfur hexafluoride. The last three are industrially produced gases.

The Kyoto meeting has also become famous for the arrival of US vice-president, Al Gore, in the last week of the negotiations. Gore, who was later to win the Nobel Peace Prize together with the IPCC, argued for a strong reduction in greenhouse gas emissions. Yet, Gore's hands were tied by a resolution passed by the US Senate, known as the Byrd–Hagel Resolution, named after the two senators who proposed it. The core of this resolution would kill the Kyoto Protocol at birth because it stipulated that the US would not be a signatory to any UNFCCC resolution that would mandate commitments to Annex-I countries, without non-Annex-I countries also agreeing to some commitments themselves. The Senate had passed the resolution with a 95–0 majority! Al Gore was thus arguing for a strong agreement, while knowing that the protocol would not stand a chance of being ratified by the US senate. A situation that presents something of a conundrum. There is in fact a strong suggestion that these tactics allowed the Clinton–Gore administration to present a climate-friendly face, without committing the US to strong reductions in emissions or the implementation of a carbon tax, something which would never have passed the US senate. Cynical as this may sound, it is not an uncommon negotiation tactic, and this presumption received support among a number of participants in Kyoto that were interviewed after the event and asked to comment on the issue.[10] So, while the US negotiated a strong reduction in emissions after Gore's arrival, this was mainly to serve their longer-term goal of presenting a climate-friendly face for the Clinton administration. It may also have helped Al Gore's standing on the issue for his own presidential bid a few years later.

In Kyoto the developed countries finally agreed to limit their emissions by 5.2% on aggregate during the period 2008–2012. The US agreed to a reduction of 7% below the 1990 benchmark, a commitment that is widely thought to have come from Al Gore's insistence after his arrival to show more flexibility in the negotiations. Before his arrival the US had held firmly to the position that they accepted no reduction compared to 1990, a position that would force them to reduce to the 1990 levels anyway since their real 1990–1995 level was already 7% higher than 1990. The EU had originally pushed for a 15% reduction but finally settled for 8%. Apart from the reduction commitments, the Kyoto protocol set out several other important elements that allowed Annex-I countries to achieve some of the reductions abroad. These were Joint Implementation, the Clean Development Mechanism, and emissions trading. Under Joint Implementation, countries with commitments under the Kyoto Protocol could transfer and/or acquire emission reduction units and use them to meet part of their emission reduction target.[11] This thus applied only to

---

[10] Hovi, J., Sprinz, D., & Bang, G., 2010
[11] See UNEP. See also van der Gaast, W., 2017 and Grubb, M., et al., 1999.

developed countries that were part of Annex-I. Importantly, these emission reductions would be supplemental to domestic actions so as not to provide a backway for not reducing emissions in the country itself. Another Kyoto innovation, the Clean Development Mechanism, in contrast, deals with project-based greenhouse gas emission reduction as a cooperation between developed and developing countries. Here emission reductions achieved through Clean Development Mechanism projects could be transferred to developed countries. The total number of registered Clean Development Mechanism projects amounted to 7859 in June 2021. Finally, emissions trading allows countries with greenhouse gas emissions below their assigned amount under the Kyoto protocol to sell these surpluses to other countries with emissions over their assigned amount.

As said, the Kyoto agreement was dead on arrival, with the US never intending to ratify the Protocol in the Senate. In fact, several years later after George W. Bush had beaten Al Gore narrowly in the elections and succeeded Clinton as president, in March 2001 he announced that the US would not implement the protocol at all but would rather seek to increase the greenhouse gas efficiency or reduce their intensity so that emissions could continue to grow with a growing economy, albeit at a lower rate of increase than before. The role of large oil companies such as Exxon in this decision and their unsavoury misinformation campaign have been shown[12] by many books, papers, and newspaper articles. Bush also dismissed the scientific evidence claiming: *'the incomplete state of scientific knowledge of the causes of, and solutions to,*

---

[12] E.g. Supran, G. & Oreskes, N., 2017

global climate change and the lack of commercially available technologies for removing and storing carbon dioxide' prevented strong action on reductions.[13]

It would produce rather dull reading to describe all the COP meetings from COP 1 to COP 26 in detail—in fact the Earth Negotiations Bulletin series presents an excellent factual account[14]—so we can fast forward to what can be considered the key meetings. Each of these meetings also represents a step change in the evolution of the negotiations. While for the Kyoto meeting, the big historical emitters in the US and EU were supposed to share most of the burden with the distinction between Annex-I and non-Annex-I countries, in Copenhagen emphasis was shifted away from the EU–US axis towards the role developing countries were playing. Finally, at the Paris conference, a full comprehensive agreement was reached with a firm goal (1.5 °C) and shared responsibilities with different levels of ambition in the reductions for countries within a legally binding framework and including elements of civil society and NGOs.

Let us now look at the 2009 Copenhagen meeting—the next big meeting after Kyoto. It was an important meeting because the commitment period of the Kyoto protocol ended in 2012 and thus a new agreement needed to be negotiated that would start at that time with goals stretching towards 2020. The path to achieve that was set up at COP 13 in Bali, Indonesia and further preparatory steps were planned to be drawn up at Poznan, Poland (COP 14). The Bali Action Plan or Roadmap identified four themes on which countries should cooperate: mitigation, adaptation, finance, and

[13] Douglas, J. & Revkin, A., 2001
[14] Earth Negotiations Bulletin (ENB) is an independent reporting service on United Nations environment and development negotiations, https://enb.iisd.org.

technology. These were, as ever, the critical questions being negotiated: how far will countries go to limit their emissions; how
can countries, particular developing countries, adapt to climate
change; and importantly, what finance and technological mechanisms are available to achieve these? While the Bali meeting was
at least successful in setting up a timeline, the Poznan meeting
was generally deemed to have made little progress.

It thus all came down to Copenhagen to finalize the agreements for the next commitment period. Over a hundred world
leaders attended the COP ministerial meeting, making it one
of the largest such meetings of world leaders of the UN outside New York at the time. A further 40,000 people applied
for accreditation at the conference, representing governments,
non-governmental organizations, intergovernmental organizations, faith-based organizations, media, and UN agencies.[15] It was
presented as a make-or-break meeting to 'seal a deal'. It failed.
Amidst strong disagreement between countries and resentment
against the way the Danish presidency handled the negotiations,
it turned out not to be the 'Hopenhagen' many people had hoped
for. Yvo de Boer, the Dutch UNFCCC executive secretary, had
clearly stated the goals in advance of the meeting: ambition on the
mid-term emission reductions by developed countries, clarity on
mitigation actions by these countries and agreements on finance
and governance. The classic four issues that define all negotiations
at COPs, past, present, and future.

Copenhagen did not deliver on those issues and ended in turmoil. Normally, civil servants negotiate a text in the first week,
after which the heads of state fly in to finalize negotiations. Just
before this high-level part of the meeting started, the COP president

Hedegard handed in her resignation to the executive secretary of the UNFCCC. The Danish prime minister Rasmussen replaced her. This was widely seen as a failure of the Danish presidency to conclude the negotiations, even though the official reason given was that the change was appropriate given the arrival of so many (115) heads of state. Earlier on in the conference, a Danish text appeared that was leaked to the UK newspaper, *The Guardian*. Particularly developing nations were upset by the procedure bypassing the negotiation process, because this infamous text was apparently prepared in a pre-COP meeting a month before the actual start of the conference. In doing so it ignored the formal process where negotiation takes place in the working groups under the UNFCCC. In the words of a delegate: '*Those writing the "Danish text" are not adequately familiar with the process. ... You cannot just assume you understand these enormously complex issues and come up with something out of the blue. You should listen and take advice from those who know how this process works.*'[16] While, certainly with the benefit of hindsight, the Danish presidency did not act in the most transparent, or diplomatically astute way, the official negotiation groups also did not produce clear documents on which the heads of state could decide. This left them having to negotiate almost from scratch. Conflicts between developing and developed countries kept frustrating the proceedings. At one stage the G77 group, an association of developing countries in the UN, and China proposed to suspend the negotiations unless the developed countries in Annex-I of the Kyoto Protocol agreed to new emission reductions after 2012. This was perceived by many developed countries as a 'walk out' and frustrated the negotiations considerably.

---

[16] Earth Negotiations Bulletin, 2009-2

Two British newspaper journalists on site, John Vidal and Jonathan Watts from *The Guardian* and its sister Sunday paper *The Observer*, succinctly summarized the meeting: '*In the end it came down to frantic horse trading between politicians. After two weeks of high politics and low cunning that pitted the world leaders against each other and threw up extraordinary alliances between states, agreement was finally reached yesterday on an accord to tackle global warming. But the bitterness and recriminations that bedevilled the talks threaten to spread as environmental activists and scientists react to what many see as a flawed deal.*'[17] They qualified the negotiation process as a shambles and as 'negotiation by leaks' because many different texts were leaked to the media before being seen or discussed by the national negotiators. At one particular moment as many as eight different versions were being circulated! The outcome of the meeting is known as the Copenhagen Accord. This was neither accepted nor declined by the countries, but in the end simply 'noted'. Large groups of countries were unhappy with the negotiation process that led to the accord being between just the US and China. A 2 °C warming limit was set, but it did not contain the emission reductions that were needed to achieve that limit. While from the EU, the German chancellor Angela Merkel at one stage proposed an 80% reduction in emissions by developed nations by 2050. This was not agreed by China. The Chinese did not like being held to explicit numbers, partly because they could be counted as a developed nation by that time (see also in chapter 8 how China overtook the US in absolute emissions). More disappointing was the fact that the Copenhagen Accord, even in its watered-down version, was not legally binding, much to the regret of the non-governmental

---

[17] Vidal, J. & Watts, J., 2009

organizations that were present in large numbers in Copenhagen. On the positive side, the accord mentioned the need for adequate monitoring and verification of reductions, and a promise of funding at the level of US$30 billion for adaptation and mitigation in developing countries. So, while some positive progress emerged from the Copenhagen meeting, the goals of the executive secretary of UNFCCC set at the beginning of the meeting were not met at all.

The perceived failure of the Copenhagen meeting had implications for the next big meeting, COP 21, planned for Paris in 2015. The French were determined to avoid a similar failure. The French presidency had rallied the full resources of their diplomatic service to manage the negotiation process well in advance. They had named a special ambassador for the conference, Laurence Tubiana, who not only travelled tirelessly in the months leading up to the conference between parties, but also changed the dynamics of the conference by suggesting to hold the high-level meeting with heads of state first and negotiating the details afterwards. French embassies around the world were briefed about the climate crisis and the upcoming negotiations, so that they could act as intermediaries between the countries and the conference presidency. In her own words: *'The objective was clear to me. This COP, this moment, was about all of these actors' visions—states, cities, regions, civil society, faith movements, business, and finance—converging to form and reinforce a new and inevitable reality: that the economy is evolving, low-carbon development is the only solution for a safe future, and this transformation is already under way. A successful Paris outcome would be the translation of that shared understanding; an outcome which all of these constituencies could take as their own and would influence expectations in the long-term.*"[18] This broad perspective, involving stakeholders with widely diverging goals and backgrounds led to a remarkably inclusive process that basically made all participants happy. The famous French cuisine apparently did the rest.[19]

In Kyoto, a top-down rule-based approach was used which required a full compliance and enforcement mechanism. In contrast, in Copenhagen the first steps were taken to using a more bottom-up approach where countries were asked to set their own individual targets. The Paris Agreement further developed this perspective and worked towards a more 'pledge and review' approach. To be able to achieve such a rather fundamental change in governance of the treaty, the French made sure that the negotiations were highly transparent and inclusive. This ensured that the final text was indeed a 'true' reflection of the views of the parties. Almost every party praised the French presidency, and it is likely that one of the reasons for that was, that all parties felt they had been heard and consulted,[20] something that went wrong in Copenhagen.

At the centre of the Paris Agreement is a five-year cycle of submissions of plans, reviews, and increasing ambition. Countries submit their nationally determined contribution (NDC) in which they spell out their targets and ways to achieve those. Each country can specify their targets and pledges according to their own specific situation. This allows the lesser developed countries a certain flexibility to have different targets to the developed ones, while at the same time ensuring that everybody is on board. This removed one of the big stumbling blocks of the Kyoto Protocol that differentiated between countries that were bound to reduce their emissions

[18] Tubiana, L., 2021
[19] Robert, A., 2015
[20] Earth Negotiations Bulletin, 2015

(Annex-I countries) and those that were not (non-Annex-1). Each new NDC must be more ambitious than the previous one, and midway between the five-year cycles, a global stocktake takes place that assesses progress on the collective efforts on mitigation and adaptation. The idea was that through such a cycle the combined efforts of all countries would limit the global temperature rise to no more than 1.5 °C and certainly below 2 °C. A transparent framework in which it is clearly visible what countries promise and what they do, ensures compliance—albeit in a far less strict manner than the Kyoto protocol. The Paris Agreement ended up as something in between a completely bottom-up 'pledge and review' protocol and completely top-down (Kyoto-like protocol).[21] In the Paris Agreement the procedure is legally binding, not whether individual countries achieve their NDC goals. This makes it somewhat less restrictive, more flexible and freer, but at the same time requires a common understanding of the need to reduce emissions, which was always key to the UNFCCC being framed as a UN convention. If the countries do not all contribute to this joint aim, the agreement falls apart.

To some extent this falling apart happened when Donald Trump decided to withdraw from the Paris Agreement on 1 June 2017, declaring that the agreement would put *intolerable burdens*' on the American economy. The withdrawal process took one year and was immediately reversed when Joe Biden became president in 2021, but it does show the fragile nature of the agreement that essentially requires all countries to cooperate. When large countries or blocks of countries refuse to participate, there is an immediate problem in reaching the emission reduction and

---

[21] Earth Negotiations Bulletin, 2015

climate goals. It comes down eventually to the credibility of the transparency framework and the global stocktake, and how well the ambitions, that are central to the agreement, can be increased. The Paris Agreement came into force on 4 November 2016 with the first global stocktake scheduled for 2023. It involves 196 countries. The transparency framework that will analyse how well countries live up to their ambitions as expressed in their NDCs, will come into force only in 2024. Overall COP 21 in Paris was considered a major success and a landmark in providing common ambitions across countries that dealt with issues of differential capabilities and equity. A financial mechanism was drawn up whereby developed countries would assist less developed countries in achieving their goals. Importantly also, it was a legally binding document, unlike the non-legally binding Copenhagen declaration.

The next task was implementing the agreement. This was achieved in developing the Paris Agreement Work Programme during COP meetings in Marrakech, Bonn, and finally in Katowice in 2018. This involved how to deal with the NDCs; adaptation, that was an important component of the Paris Agreement and was likely one of the mechanisms for bringing developed and less developed countries together; the transparency framework and finances; and the global stocktake. In Katowice this was accepted as the Katowice Climate Package. In Paris, the COP had also requested IPCC to make a special report on how to remain below 1.5 °C and to assess the implications of going beyond that. The report when published in 2018 made chilling reading on how close the world was to reaching the limit. At Katowice, the countries, however, could not agree whether to 'welcome' or 'note' the report—resistance to welcoming the report not unexpectedly came from major oil- and gas-producing countries such as Saudi Arabia, the US, Russia, and Kuwait. The main achievement of COP24 was its 97-page guidebook containing the operating guidelines for implementing the Paris Agreement.

COP25 was held in Madrid, after civil unrest made it impossible to hold it in Peru, which had been selected after Brazil, under its right-wing president Jair Bolsonaro, pulled out of the organization. It was the last of the Kyoto era COPs and only eleven months away from the Paris Agreement coming into force in 2020. Madrid was seen by many as a failed conference, even though it managed some progress in the areas of 'loss and damage', harm caused by climate change that is irreversible and to which countries cannot adapt. At the same time public opinion had become stronger and stronger, stimulated by ever-increasing media attention. Climate strikes involving young people, such as the Swedish activist Greta Thunberg became common and during the conference an estimated half a million people demonstrated in the centre of Madrid, demanding more ambition from the negotiators. There was also considerable unease about the role not only of the US but also of the EU and China *as with the Kyoto Protocol before it, there will again be a climate agreement, one designed around US demands, without the participation of the world's highest per capita emitter. This raised questions of leadership, and serious concerns if the EU, hampered internally by some member states reluctant to engage on climate, or China, focused on its own development, can credibly lead the way to a more climate ambitious future.'[22]*

Then, COVID-19 struck the world. Two weeks after the closure of the conference in Madrid, the world learned about the

---

outbreak of a new virus in Wuhan, China; a virus that caused severe respiratory problems and sometimes deadly pneumonia. It was declared a pandemic by the World Health Organization in March 2020. Since then, a staggering 5.66 million people have died worldwide from the disease.[23] Amidst this turmoil, countries went into sometimes draconian lockdowns, people working from home, bars and restaurants closed, and social activities put to almost zero levels. In the next chapter we will analyse some of the impacts of these lockdowns on emissions. The impact on the UNFCCC process was severe. For the COP meetings it was impossible to convene physically due to travel and quarantine restrictions. However, being together is almost an absolute requirement for a multi-lateral negotiation process, as digital meetings across several time zones make it very hard to reach consensus. Thus, after delaying for a year, COP 26 finally took place in December 2021 in Glasgow. The irony that the meeting was held in the town of James Watt, the inventor of the steam engine (see Chapter 8), was probably not lost on the participants. While 40,000 people had registered, either for physical presence or as an online participant, there was only room for 10,000 people at a time in the conference buildings. Meetings were held with the required social distancing and many delegates complained about not being able to enter the conference grounds or the rooms. It also highlighted similar equity issues that plagued climate change negotiations with a large divergence between the haves (those that had access to vaccines) and the have nots, the poorer countries that did not have equal access to vaccines. This complicated the already difficult agenda even further.

[23] Our World in Data, 2022

COP26 was the first COP after the Kyoto Protocol formally came to an end. It had the difficult but clear task to show progress in battling the climate emergency. In August 2021 the IPCC had published its Sixth Assessment report, stating that '*It is unequivocal that human influence has warmed the atmosphere, ocean and land. … Observed increases in well-mixed greenhouse gas (GHG) concentrations since around 1750 are unequivocally caused by human activities.*'[24] If there was still any doubt in the minds of the delegates about the impact of humans on the planet's climate, the report did everything to quell it. The report also made abundantly clear that the world was on course to go beyond 1.5 °C in a few years if no immediate action was taken. The gigantic task to keep the 1.5 °C promise of the Paris Agreement alive fell heavily on the shoulders of the UK chair of the presidency Alok Sharma. Public pressure for noticeable progress was immense and had grown with the increasing level of extreme climate events that were now afflicting the Earth. Greta Thunberg, the Swedish schoolgirl that had initiated the Friday climate strikes, by now had achieved global stardom and had millions of youths demonstrating for actions on climate change. The UN's own Environment Programme had calculated in its emissions gap report that if the world followed the measures promised in the countries' NDCs the world would warm by 2.7 °C by the end of the century, way beyond the Paris target of 1.5 °C.[25] Clearly more ambition was needed, and this led to calls to speed up the five-year cycle and for countries to adapt their NDCs more quickly to changing circumstances.

In the first week the delegates largely ignored the elephant in the room, i.e., the reduction of CO_2 emissions through the burning of

---

[24] IPCC, 2021-2
[25] United Nations Environment Programme, 2021

fossil fuels. They did agree, however, on halting and reversing forest loss through deforestation and land degradation by 2030 and on making 75% of the wood products sourced from sustainable forests. This Glasgow Leaders Declaration on forest and land use was signed by 120 countries, including Brazil and Indonesia, two of the countries with the highest deforestation rates. The Leaders also pledged to reduce methane emissions by 30% by 2030 in the Methane Pledge, signed by 100 countries amongst which were the US and the EU, but not, for instance, Russia. China and the US brokered several joint deals including the methane pledge and a promise to accelerate the reduction of the use of coal. This however led to a standoff at the final meeting between India, China, and a group of countries like Switzerland and the EU that wanted to see a clear ambition to 'phase out' coal use altogether. In the end the dispute and the request by India, supported by China and the US to change the wording into 'phase down', brought the president Alok Sharma close to tears. It was one of the many compromises of the Glasgow pact, a pact that left most delegates dissatisfied with the overall text in which they found little wins but considerable losses. COP 26 was also not able to agree on a 100-billion-dollar fund from developed nations to assist less developed countries in adaptation and battling climate change, this of course causing much resentment among nations that are already suffering badly from climate change. The closing speech of the UN Secretary General, António Guterres, summed it up: *The approved texts are a compromise. They reflect the interests, the conditions, the contradictions and the state of political will in the world today. They take important steps, but unfortunately the collective political will was not enough to overcome some deep contradictions. Our fragile planet is hanging by a thread. We are still knocking on the door of climate catastrophe. I reaffirm my conviction that we must end fossil fuels subsidies. Phase out coal. Put a price on carbon.'*[26] Guterres' analysis of the text as a compromise could have been applied to almost every COP held since Berlin, but it was the first time that fossil fuel subsidies and the phasing out of coal were seriously on the agenda. Paraphrasing Neil Armstrong, this may be a small step for now, but it can turn out to be a giant leap for humankind.

[26] https://www.un.org/sg/en/content/sg/statement/2021-11-13/secretary-generals-statement-the-conclusion-of-the-un-climate-change-conference-cop26

CHAPTER 12

THE BUMPY ROAD TO THE FUTURE...

We try to look at the future. At the end of 2021 and the beginning of 2022, the IPCC released its three reports on the state of the climate, the impacts and mitigation. After thirty, or maybe fifty, years the message could not be clearer. The world was warming more than 1.1 °C, and the world would likely miss its 1.5 °C target if it did not drastically reduce its emissions. Then the war in Ukraine started and the dependence of the western world on (Russian) fossil fuel became painfully clear. The world indeed was addicted to fossil fuel and it is likely that we are on course for a devastating 3 degrees warming. This caused concerned and frustrated scientists to take to the streets and call for action. A desperate call that was supported by the UN secretary general in clear words: 'Climate activists are sometimes depicted as dangerous radicals, but the truly dangerous radicals are the countries that are increasing the production of fossil fuels. Investing in new fossil fuels infrastructure is moral and economic madness.'

Time is running out for the world to remain below 1.5 °C or even 2 °C. To do this would require massive changes in the way we use and produce energy. We should end our dependence on coal almost immediately and subsequently very quickly reduce our dependence on oil and gas. All the scenarios that keep temperature below 1.5 °C and 2 °C require deep cuts in emissions *as from 2020, at the level slightly above the Covid-induced reduction of 5.4%. This gives an idea by how much, for instance, transport emissions must be reduced. However, the emissions recovered after the initial COVID decrease, while they needed to be sustained to achieve the overall required reduction. There is theoretically a safe landing space for our planet, in terms of per capita emissions that meets four key sustainable development goals: ending poverty, zero hunger, good health and wellbeing, and clean energy. Achieving this requires developed countries to immediately change their technologies and practices, while the less developed nations need to make sure that they do not embark on high-emission technologies while building new wealth and enhancing wellbeing.*

That transition is still possible through the increased use of renewables: wind and solar energy. Large reductions can be achieved by massive deployment of these at lower costs than other sources and there are signs that renewables are indeed increasingly being used. Reliance on unproven technologies such as Carbon Dioxide Removal to release the burden on immediate reductions puts the burden on future generations. Now, early 2022, is the time to reduce our dependence on coal, oil, and gas almost immediately. This requires huge transitions, not only in our way of living, but also in our economies. Novel economies should promote social welfare without breaking ecological and planetary boundaries.

Just 370 years after the discovery of carbon dioxide as a wild spirit, by the Flemish medical doctor Jean Baptista van Helmont, we are hitting 420 ppm (summer 2022), way above the normal range for the Pleistocene, and 50% more than the pre-industrial level. At current emission rates, it will take about 8–10 years before we have burnt through our carbon budget for 1.5 °C. With every tenth of a degree further warming the impacts will be larger and more severely felt. We need everyone and every tool to help us

in reducing emissions before carbon dioxide becomes truly van Helmont's
wild spirit and puts our environment on a direct path to catastrophe. It can
be done...

On 4 April 2022, the IPCC released its report from Working
Group III, Mitigation of Climate Change. While the release of the
Working Group II report about a month earlier on the impacts
of climate change coincided with the start of the Ukrainian war
and drew little attention, the Working Group III report featured
more prominently in the media. Concerned and anxious activists
united under the Scientist Rebellion group demonstrated globally
in more than 25 cities and chained themselves to a range of institutions such as, for instance, the US bank JP Morgan. These activists
were exhausted and frustrated by the continued dismissal, ignorance, and outright denial of climate change by governments
and big financial institutions that continue to support fossil
fuel extraction by energy companies. The UN secretary General
António Guterres stated in a press conference at the release of the
IPCC Working Group III report, in as strong a language as ever
heard from the world's top diplomat, *'Climate activists are sometimes*
depicted as dangerous radicals, but the truly dangerous radicals are the
countries that are increasing the production of fossil fuels. Investing in new
fossil fuels infrastructure is moral and economic madness.'[1]

The US government announced after the Russian invasion that
it would look again at fracking to supply gas as banning imports
from Russia was increasingly being discussed. The brutal invasion
of Ukraine by Russia at the end of February 2022 not only showed

[1] United Nations, Secretary General, 2022

the extent to which the Western world is addicted to fossil fuel, but also how severely poisoned it has become by this addiction. The Western world continues to buy gas and oil from Russia, despite the severest regime of economic sanctions ever agreed upon. Anything, but our oil and gas.... Russia, on the other hand, still pays off its international debts. Conforming to international financial regulation trumps behaving barbarously on the battlefield. Oil, gas, and money are intricately linked in today's economic and political world.[2]

Part of that politics plays out every year at the COP as the difference between 'words and deeds'. While the European Commission's vice-president Frans Timmermans expressed almost emotional support for the inclusion of phasing out the use of coal in the final agreement, many countries complained that the practice within the European Union was still pointing in a rather different direction. Saying one thing and doing another is easier in a Paris Agreement pledge and review system which only demands moral transparency than in a Kyoto compliance regime where bad behaviour is meant to be punished. A similar pattern of behaviour can be observed in several countries, such as Germany, that have become strongly dependent on the supply of Russian gas and still (at the time of writing, early 2022) refuse to cut down their dependency in response to the Russian invasion—fearing the economic consequences. In the Netherlands, the government decided to reduce tax on petrol so that people could continue to drive, despite sky rocketing prices, rather than embark on a

---

[2] See the publications of the Swedish human ecologist Malm, A., in particular Fossil Capital, 2016, and the more recent Malm, A., & the Zetkin collective, 2021, in which they explore the links between climate denial and the ultra-right & anti-immigration movements.

strategy to get as many people as possible to travel by public transport or by electric bikes or vehicles. Even when the government decided to recommend lowering the thermostats in houses and official buildings, the main argument used was not to save on fossil fuel and hence CO_2 emission but saving on the expense because of the war in Ukraine.[3]

There are, however, also some hopeful signs of increased action among not only civil society such as non-governmental organizations, and local city governments, but also manufacturers, particularly as public awareness and outrage have increased since the Copenhagen COP meeting (see Chapter 11). This local-level attention may mean that societies are indeed on their way to transforming their energy systems. Overall, the EU's Green Deal is an unprecedented programme of stimulating green energy development, changing agriculture and transport that would bring the Union to net zero in 2050. In the US the Biden administration has come up with similar plans, although they face a divided senate. China also continues to move towards more renewable energy sources. Seventy-eight countries, including the EU27, have also pledged to reach net zero emissions by 2050. This is the minimum needed at the global level to remain below 1.5 °C.

What is this 'net zero' that has started to dominate the discussion on climate change mitigation? Net zero means that we have cut greenhouse gas emissions to as close to zero as possible. The small amount of necessary remaining emissions should then be re-absorbed by oceans and land. It arises from the concept of an allowable carbon budget we discussed in Chapter 10. Sanctions have also severely reduced the dependence on Russian oil and gas.

---

[3] Developments in the energy reliance and gas dependence of Europe from Russia are moving very fast and at the time of writing, July 2022, Russia appears to be cutting down gas exports to Germany by more than 60%.

Given that there is such a strong relation between temperature increase and cumulative emissions (the TCRE curve), it is also clear that when one reaches a certain cumulative amount—and emissions would need to stop to not further increase the cumulative emissions—the temperature will also not increase further. This temperature will remain relatively stable for decades to a few centuries. The physical background is that the uptake of CO_2 by the deep ocean from the atmosphere acts to cool the Earth but is balanced by the release of heat from the deep ocean which warms the planet (heat absorbed since 1850 by the oceans). These processes are in balance on the timescales of decades to centuries, no longer or shorter.[4] Net zero also applies to the whole globe, as this is what the curve that relates cumulative emissions to temperature is based on. Thus, while individual countries might claim to move towards net zero, this is not sufficient for the overall temperature to remain below the 1.5 °C target. This requires the whole global community to reach net zero on average. Also, note that net zero also includes the non-CO_2 greenhouse gases of N_2O and CH_4, and that, for the time being, we ignore these, as reducing CO_2 is what really matters in the long term.

At the time of writing, July 2022, all the reports of the three working groups of the IPCC 6th assessment cycle have been published. They all contain dire warnings about the state of the climate and how close we are to missing the 1.5 °C target of the Paris Agreement. Where are we, after the Glasgow meeting and the three IPCC reports? Does our future indeed look bleak? The short answer is yes. We are still far from reaching the Paris ambitions on temperature (1.5 °C) if we sum up the expected contributions of the countries to emission reductions as expressed in their

---

[4] Fankhauser, S., et al., 2021

NDCs. Furthermore, since the start of the observations by Charles Keeling in 1958, the growth rates of the atmospheric CO_2 concentration have only increased: from 1 ppm per year in the early 1960s to 1.5 ppm per year in the 1980s to 2 ppm per year in 2000–2005 to about 2.5 ppm per year at present. 17% of the cumulative emissions took place between 2010 and 2019, 42% between 1990 and 2019, again strongly increasing rates. These increases continue to occur, despite the 26 meetings of the Conferences of the Parties.

On the other hand, it has become increasingly clear that it is unlikely that the world will end up in the high-end scenarios of the IPCC that lead to 4 °C warming, or even more. The balance between oil and gas reserves, the increasing cost of exploitation, and, importantly, the decreasing costs of renewables makes it unlikely that these dramatic scenarios will become reality. However, as Figure 12.1 shows, we are still on course for a 3 °C warming with all its dramatic impacts if we follow our current policies and only to a 2 °C warming if we follow national pledges; 1.5 °C is going to be very difficult. While the likely failure of reaching 1.5 °C is, of course, disappointing, all is not lost for a 2 °C world. In fact, a recent publication suggested that this was possible provided all the existing pledges to reduce emissions were implemented fully and on time.[5]

One of the curves in Figure 12.1 projects a path towards 1.5 °C. It features something quite important towards the end of this century: it dives below the zero-horizontal line of emissions. This means that by that time we should be taking CO_2 out of the air to reach the 1.5 °C target. This is called negative emission, and the technologies to achieve this are known as negative

[5] Meinshausen, M., et al., 2022

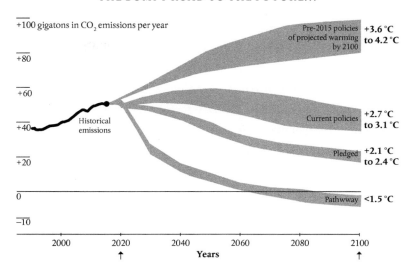

Figure 12.1 Historical fossil-fuel emissions up to 2020 and future fossil-fuel emissions that would result in the CO_2 concentrations in 4 IPCC scenarios. Drawing based on United Nations Emissions Gap report.
United Nations Environment Programme, 2021

emission technologies (NET) or Carbon Dioxide Removal (CDR). The simple explanation is that the cumulative amount of CO_2 in the atmosphere has become too high to remain within the temperature target. To understand why CDR is needed we need to a go a little into how these scenarios are constructed. The tools with which these scenarios are made are so called Integrated Assessment Models.[6] These are models that are based on a particular future or storyline of a future; they make predictions of the emissions resulting from those assumptions in the storyline. These emissions are then fed into simplified climate

[6] van Beek, L., et al., 2020

models to obtain a temperature scenario for that future. Integrated assessment models combine insights from climate science and economics to estimate how industrial and agricultural processes might be transformed to tackle global warming. They contain model assumptions about technologies, such as high-emission power plants being replaced by emission-free ones, and the cost of electric vehicles, and economic processes. To take the two examples used above, they enable us to investigate how a carbon tax might help to achieve cuts in emissions, or decarbonization of the transport sector could shift investments towards greener fuels and electricity.[7] They are based, like most existing economic theories, on rational choices societies could make. Of course, sometimes the choices made are not rational—for example, the impact of the Ukrainian war on fossil fuel use and prices. They also contain highly simplified descriptions of the interaction of humans with land use and technology. They are also notoriously poor at incorporating the effects of policies and changing governance.[8] Nevertheless, these models have become one of the main tools to sketch out alternatives to climate mitigation and are central in the IPCC analyses.

IPCC groups the different scenarios of these models into several classes. In the latest report they distinguished eight classes roughly based on the temperature targets shown in Figure 12.1. In each scenario a particular set of technologies is implemented and their interaction within the economy and population calculated in terms of emissions. If such a scenario then ends up at, say, the target of 1.5 °C in 2050, the measures used in that scenario are considered necessary to reach the target. However, a different

[7] Peng, W., et al., 2021
[8] Peng, W., et al., 2021

set of technologies and measures might also lead to the desired outcome, so there is in fact a multitude of pathways to reach the desired target. This gives the band width of the lines in Figure 12.1 for each temperature pathway. Technologies that are common to all the scenarios for a particular target are deemed essential to reach it. So, if the majority of 2 °C class scenarios contain a move away from heavy fossil fuel use in power generation, the IPCC reports states this as ... *'In modelled global pathways that limit warming to 2 °C (>67%) or lower, most remaining fossil fuel CO_2 emissions until the time of global net zero CO_2 emissions are projected to occur outside the power sector, mainly in industry and transport. Decommissioning and reduced utilization of existing fossil fuel-based power sector infrastructure, retrofitting existing installations with CCS switches to low carbon fuels, and cancellation of new coal installations without CCS are major options that can contribute to aligning future CO_2 emissions from the power sector with emissions in the assessed global modelled least-cost pathways.'*[9] CCS in the previous quote stands for carbon capture and storage, a technique whereby the CO_2 in the exhaust is captured and stored, mostly underground in geological reservoirs. The use of CCS makes it theoretically possible to generate energy at almost emission-neutral levels (or at least 90–95%), but its efficiency is controversial.

All the scenarios that keep temperature below 1.5 °C and 2 °C require deep cuts in emissions as from 2020,[10] at the level slightly above the Covid-induced reduction of 5.4%. The timing at which net zero is reached, however, differs between the scenarios. 1.5 °C scenarios generally reach it on average between 2050 and 2055;

---

[9] IPCC, 2022-1
[10] It is interesting to note that the key reductions in the emission pathways of IPCC start in 2020. In 2022 we have not yet seen any of these reductions in reality. This reduces the time we have for reductions by another 2 years, compared to 2020.

2 °C scenarios add an additional 10–20 years, and scenarios that follow current policies and NDCs reach it around 2070. This should not come as a surprise—depending on the target temperature, the carbon budget allowed is slightly larger, so there is more time to burn fossil fuel before the net zero point is reached. However, these lower range scenarios also use a considerable amount of negative emission or active carbon removal from the atmosphere. Some of the scenarios do without this condition, but in fact most must use them to achieve their target. This is what we see in Figure 12.1 for the 1.5 °C pathway.

Negative emission technologies comprise a wide range of techniques that have various stages of what is called technological readiness (how quickly they can be applied, and at the required scale). They include afforestation and reforestation (trees take up carbon as we know, Chapter 5), bioenergy with carbon capture and storage (BECCS), direct air capture and storage (DACCS), enhanced weathering (see Chapter 4), biochar (pyrolysis, burning of organic material without oxygen, and adding it to the soil where it improves the structure and remains stable), wetland restoration, and ocean alkalinity treatment (see Chapter 9). None of the techniques mentioned has shown to be applicable at the scale required (say removing up to 2 Gigatonnes of carbon per year, one fifth of our current emissions) and none of the scenarios that use CDR avoid the need for deep emission cuts in the twenty-first century. Thus, the largest problem with relying on CDRs is that their application could obstruct, halt, or reduce the required emission reduction effort in the short term. Take for instance the analysis of IPCC: '*the prospect of large-scale CDR could...obstruct near-term emission reduction efforts, mask insufficient policy interventions, might lead to an overreliance on technologies that are still in their infancy, could overburden future generations, might evoke new conflicts over equitable burden-sharing, [and] could impact food security, biodiversity or land rights.*[11] Virtually all the techniques themselves have problems: the land-use techniques require land, land that is currently used for agriculture, the ocean and weathering techniques require upscaling and rigorous testing for side effects, while DACCS requires energy which sort of beats its own purpose, unless that energy is produced sustainably (but even then, the amount is huge), and so on and so on. A recent analysis therefore concluded the reliance on unproven future technologies for CDR is inappropriate.[12] Reliance on Carbon Dioxide Removal puts the burden on future generations. That is not to say that we should not investigate its use and explore its potential. In contrast, every reduction helps, and one could argue that a return in the future to CO_2 concentration levels that approach pre-industrial levels may be the best long-term strategy for the planet's conservation. But in the short term we simply need to cut our emissions, fast and drastically.

Figure 12.1 shows the immensity of the task ahead. After years of relentless growth in CO_2 emissions to the atmosphere, since 1850 in fact, we need to slow down the rate of increase and touch the zero level of emissions around 2050 to remain below 1.5 °C. This is what the current generation of Earth System models tells us. The authors of a blog post in Carbonbrief discuss this further. They suggest that '...*if the rate of rising atmospheric CO_2 does not begin to slow markedly within the next few years, the chances of limiting global warming to 1.5 °C will rapidly vanish.*'[13] This statement was reflected in the

---

[11] IPCC, 2022-2
[12] Calverley, D. & Anderson, K., 2022
[13] Betts, R., et al., 2022

report of Working Group II of the IPCC that deals with impacts, adaptation, and vulnerability: '*There is a rapidly narrowing window of opportunity to enable climate resilient development.*'[14] The IPCC Working Group III report echoed this statement in various other forms. Stopping the warming is thus physically possible, but whether we can move away from our fossil fuel addiction quickly enough, to turn a still increasing use into a decreasing use is not straightforward. Time is running out. Just look at the numbers in a bit more detail. To have a 50% chance of limiting global warming to 1.5 °C, the TCRE curve suggests an allowable budget of 120 Gt C, for 1.7 °C this is 210 Gt C, and for 2 °C this is 350 Gt C. If we assume 2021 emission levels these budgets are reached in 11, 20, and 32 years. And remember, if we want a higher probability of achieving the target, these numbers must be even smaller, and the time to reach them even shorter. There are not many of us that would be happy with boarding a plane when the chances of it falling out of the sky are fifty-fifty. Yet, this is what a 50% probability means, it is the same as tossing a coin.

Covid lockdowns in 2020 led to a reduction in fossil fuel emissions of about 5.4%.[15] However, to reach the low Paris 1.5 °C target, such reductions need to be sustained, continuously, not just for a single year. It is not a comforting thought to realize that after the initial reduction in emissions, a rebound took place, very similar to previous financial crises, such as in 2008, that caused temporary reductions in emissions. This rebound was estimated to be around 5% as well. The reduction of 5% was too small to be noticed in a decrease in the CO_2 growth rate that remained at 2.3 ppm year^{-1}.

---

[14] IPCC, 2022-1
[15] Friedlingstein, P., et al., 2020

The current pathway of our emissions thus puts the lowest Paris target virtually out of reach. The most recent UN Emissions Gap report, prepared for the Glasgow meeting, was already painfully clear about this: *'The emissions gap[16] remains large: compared to previous unconditional NDCs, the new pledges for 2030 reduce projected 2030 emissions by only 7.5 per cent, whereas 30 per cent is needed for 2 °C and 55 per cent is needed for 1.5 °C.'* These pledges are a far cry from what is needed. Collectively, countries are falling short in their reduction pledges, some more than others. The report also assesses the combined impact of new pledges (the NDCs). Just under 50% of the new or updated NDCs submitted to UNFCCC would result in lower emissions in 2030 than their previous NDCs. These countries are responsible for 32% of the total emissions. About 18% of the NDCs, from countries accounting for 13% of global emissions, state no reductions of their 2030 emissions relative to their previous NDC. Of the remaining NDCs it was hard to determine whether a reduction would be achieved. Argentina, Canada, the EU27, South Africa, the United Kingdom, and the United States of America have stated reductions that would collectively lower the emissions by 0.6 Gt C compared to earlier NDCs. The UN report specifically addresses the responses of G20, the 20 largest economies in the world. The G20 is of course historically mostly responsible for CO_2 emissions, but collectively they are not on track to achieve the required reductions. While the G20 countries have adopted various policies in recent years, and there are many positive developments, there are also some examples that

---

[16] The emissions gap for 2030 is defined as 'the difference between total global GHG emissions from least-cost scenarios that keep global warming to 2 °C, 1.8 °C, or 1.5 °C with varying levels of likelihood and the estimated total global GHG emissions resulting from the full implementation of the NDCs', United Nations Environment Programme, 2021.

are difficult to reconcile with these pledges. These comprise plans and projects for fossil fuel extraction and coal-fired power plants (in China) as well as rollback of environmental regulations during the COVID-19 pandemic. A complex picture is thus emerging, and one in which there are also G20 countries that have submitted NDCs that are likely to result in an increase in emissions (Australia, Indonesia).[17]

Outside the G20, there is a very unequal distribution in where countries are in their trajectory of CO_2 emission and their development level. IPCC's Working Group III developed an illustrative graphic showing where the safe landing space is for our planet, in terms of per capita emissions and the meeting of four key sustainable development goals: ending poverty, zero hunger, good health and wellbeing, and clean energy.[18] It is possible to group these into an index, as, for instance, by the historic index of human development. An index above 0.5 generally means that the development goals have been reached or are within reach; an index lower than 0.5 tells you that there remains work to do. Similarly, emissions of more than 5 tonnes CO_2 equivalent per capita (so including other greenhouse gases such as CH_4 and N_2O) would put the Paris Agreement targets out of reach. And now comes the interesting, but not wholly surprising part. Developing nations in East, West, South, and Middle Africa, and India and Sri Lanka are generally below the emission boundary but also below 0.5 on the development index. Eastern Asia and the developing Pacific

---

[17] United Nations Environment Programme, 2021

[18] 'The Sustainable Development Goals (SDGs), also known as the Global Goals, were adopted by the United Nations in 2015 as a universal call to action to end poverty, protect the planet, and ensure that by 2030 all people enjoy peace and prosperity. The 17 SDGs are integrated—they recognize that action in one area will affect outcomes in others, and that development must balance social, economic, and environmental sustainability. See United Nations Development Programme, 2022

are in contrast above the emission target, but just above the index of 0.5, with Eurasia and the Middle East reaching the 0.5 index value exactly, but at an emission of 10–15 tonnes CO_2 equivalent per capita. The developed nations group roughly in two, the European and Asian Pacific developed nations have relative high values of human development at 0.7–0.8 (the maximum is 1) but at the cost of relative high emissions, double the boundary value of 5 tonnes CO_2 equivalent per capita. The other group consists of US, Canada, New Zealand, and Australia that have high values of the development index, 0.8–0.9, but also the highest per capita emissions (20 tonnes CO_2 equivalent per capita). Most of the G20 countries 'sit' on the high side of the emissions and development index. It is not just development stage that is important as the IPCC notes—income or wealth matters too: *'Variations in regional, and national per capita emissions partly reflect different development stages, but they also vary widely at similar income levels. The 10% of households with the highest per capita emissions contribute a disproportionately large share of global household GHG emissions.'*[19]

However, to move towards a more equitable world, the countries with low emissions need to move towards a higher development index value, while the countries with high index values need to reduce their emissions. There is thus just one implication that can be drawn here, that identified by IPCC WGIII: namely that the developed countries need to rapidly change their technologies and practices to achieve massive reductions in their fossil fuel emissions, while the lesser developed nations need to make sure that they do not embark on high-emission technologies and

---

[19] IPCC, 2022-2

invest further in eradicating poverty and hunger, while building wealth and enhancing wellbeing. It is now or never, or phrased differently, the global clock is ticking. Is this at all possible, given that we have spent at least 30 years in the knowledge that climate change was real and that something needed to be done, and arguably, not much happened?

It is technologically possible to achieve the required reduction, although this requires large transformations in societies. The IPCC Working Group III report has produced an excellent graph that puts the costs of technologies against the current cost of carbon on the emission market (Figure 12.2).

They classify the technologies for energy, AFOLU (Agriculture, Forestry, and Other Land Use), buildings, transport, and industry. Most striking among those are the blue areas, where cost of mitigation, such as for the renewables wind and solar energy now are low enough to compete with the cost of carbon and can thus be implemented straight away in economic terms. Large reductions up to 2 Gt CO_2 per year could be achieved by massive deployment of these at lower costs than other sources. The cost of renewable energy, particularly solar, has continued to decrease so rapidly, that cost is in principle no longer an obstacle. This is one of the most optimistic signs that energy transformations are indeed possible and can deliver at a much faster rate that was estimated previously. The economic feasibility of some of the other options depends on the cost of carbon in the emission market. In the EU emission trading system, the world's largest emission trading system, the cost of carbon in 2022 is around US$80 per ton CO_2. Such a carbon pricing mechanism is seen by many economists as central to any decarbonization policy, among them William Nordhaus, who was awarded the 2020 Nobel Prize for

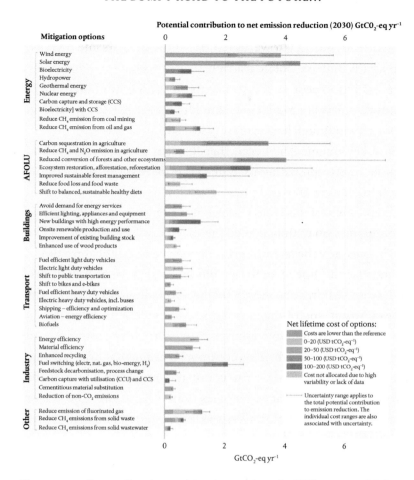

Figure 12.2 Costs of various mitigation options (in US$ per ton CO₂).
IPCC, 2022–2

Economic Sciences, for his work on the interaction of climate and economics, in particular carbon pricing.[20] It provides a benchmark and, importantly, taxes bad behaviour. The other area

---

[20] Nordhaus, W., 2013. But see also Keen, S., 2021 for a critique of the neoliberal approach of Nordhaus that underestimates the damage caused by climate change.

where reductions can be achieved relatively easily is in transport, primarily by shifting towards electric bikes and cars and increasing accessibility to public transport. Even when we accept a low rate of carbon cost at US$50 per ton CO_2, we still have quite a number of readily available technologies to make the transition from a fossil fuel-driven economy to a sustainable one.

The key issue is, however, that it needs to happen fast, and this is where governments loiter. The discussions at Glasgow about the wording of the 'phasing out' or 'phasing down' of fossil fuel are telling. The Dutch prime minister was initially against signing the declaration, protecting the interest of companies like Shell; only after severe criticism from several political parties at home and outrage by activists, did he make a U-turn and supported the declaration. Fossil fuel is deeply ingrained within the power structures of governments worldwide and it requires a fundamentally different approach to stop their subsidies and reduce their influence. The common incremental and lacklustre approach followed so far is not moving them towards the production of renewable energies. Good signs are that major pension funds, several universities, even some banks and other institutions are de-investing from fossil fuel companies. The governments of the western world, those with large historical contributions to emissions, a high human development index, and a high emission per capita, should set the lead. It is one thing to tell other countries to reduce their emissions, it is quite another to set the moral example by phasing out rapidly your own fossil fuel dependence. That takes courage, something most governments lack, concerned as they are with short-term electoral gains. But it is something the 'common but differentiated responsibilities' dogma that forms the basis of the UNFCCC calls for.

Using the allowed carbon budget, the UK Tyndall Centre conducted a similar analysis to the IPCC's, with similarly stark conclusions that provide a roadmap to phasing out (note, not phasing down!) fossil fuels.[21] The production of coal, one of the worst fossil fuels at emitting CO_2, needs to be phased out first. For developed nations this would require a reduction of 50% in the next five years, with developing nations needing to cut down their production in a further five years. This is in line with the UNFCCC's 'common but differentiated' policy goal. Oil and gas production by the wealthier nations (those generally with a high development index) needs to be reduced by 74% in the next ten years and completely in slightly more than 10 years (2034). For low- and medium-income countries these cuts should be 14% and 28% at 2030 with complete shutdown in 2050 and 2043 respectively. The maths on which this is based is remarkably simple, making the analysis not dependent on the type of assumptions used in IAMs. The amounts of the reductions required leave no room for the development of new fossil fuel extraction sites. There is simply no room in the allowable carbon budget. While the developed nations need to take the lead, the size of the carbon budget implies that developing nations should also share the burden. To make this feasible, large financial transfers are needed from the rich countries to sustain the economies of the poorer countries. Unlike the IPCC scenarios based on Integrated Assessment Models with Carbon Dioxide Removal, the Tyndall Centre report used stylized scenarios that contain no Carbon Dioxide Removal.

[21] Calverley, D. & Anderson, K., 2022

These scenarios are essentially heuristic tools that achieve the same reduction, but in slightly different ways, such as initial fast reduction and lower rates at the end, or the reverse. Since the cumulative budget is what counts, this simplifies the procedure and makes no assumptions about the precise economics, population growth, land use, or new technologies that form the basis of the assessment models. They set the goal, not the tools.

It is time to face the elephant in the room and reduce our dependence on coal, oil, and gas almost immediately. This requires huge transitions, not only in our way of living, but also in our economies. It is not impossible to do that, as Figure 12.2 shows. Importantly, this is not only a matter of individual responsibility. Most people are aware of the danger of climate change and are quite willing to change their diets, put their thermostats lower, and buy electric cars. It is the governments of the world that need to change; they need to stop subsidizing fossil fuel production. They also, urgently, need to introduce a carbon tax so that technologies that do not produce carbon dioxide become cheaper than those that do. They need to rethink the way their economies are structured, moving away from a paradigm that allows a small percentage of the world to not only become super rich, but also become responsible for the majority of the emissions. Again, this is not impossible—several cities have started to look at novel ideas such as the 'doughnut economy'[22] that aim to promote social welfare without breaking ecological and planetary boundaries. These ideas require us to shed the fossil fuel addiction. This is where individuals can play a role, primarily as activists forcing

[22] Raworth, K., 2017

our governments to go into rehab[23]: acknowledgement of having caused the problem first, and then break the habit. The world's hopes lie with these activists. They lie with Greta Thunberg who single-handedly organized the youth of the world in Friday school strikes. It lies with the scientists of Scientist Rebellion who have chained themselves to banks supporting fossil fuel investment and government buildings where politicians are housed that continue to ignore the science and continue to support fossil fuel companies.

So, this is where we are, some 370 years after the discovery of carbon dioxide as a wild sprit, *spirit sylvestris*, by the Flemish medical doctor Jean Baptista van Helmont. It took us a little less than 200 years to burn such an amount of fossil fuel that the atmospheric concentration of carbon dioxide rose from 280 to 412 ppm and the global temperature by 1.1 °C. By around 1850, the key chemistry of the burning process and the existence of a geological cycle were known. In the second half of the nineteenth century, the impact of CO_2 on the climate was established first by Eunice Foote and later by John Tyndall. Scientists like Ébelmen had started looking at the key role of the geological carbon cycle and its variability at a geological timescale was starting to be established. The crucial dance of ice in the glacial-interglacial cycles of the last million years was a discovery that triggered Arrhenius to calculate the impact of CO_2 on Earth's climate towards the end of the nineteenth century. Apart from the work of Callendar, science remained eerily silent until some 60 years ago when Keeling started to measure the rise in concentration of anthropogenic

---

[23] This is essentially the role the Canadian author Naomi Klein advocates forcefully in her books, such as '*This changes everything*'. See Klein, N., 2015.

CO_2 in the atmosphere. In 1972 at the UN conference in Stockholm, the issue of man-made climate change through burning fossil fuel was first put firmly on the agenda of the world's governments. Since 2016, we have had the Paris Agreement that calls for ever stricter ambition to reduce our emissions.

Now in 2022, do we have time left to avoid further disaster? The scientific evidence says no. At current emission rates, it will take about 8–10 years before we have burnt through our carbon budget that would allow us to stay below 1.5 °C. Unless something drastic occurs that will almost immediately stop our dependence on fossil fuel, we will fail to reach that target. The next step, staying below 2 °C, is still possible, but requires fast and massive changes to energy systems, to our economy, and our way of living. The difference between 1.5 °C and 2 °C may appear irrelevant as in the end they are just politically negotiated values, but the thresholds do have meaning. The impacts at 1.5 °C and 2 °C are not similar— on the contrary the impacts at 2 °C are much more severe than those at 1.5 °C (see Chapter 1). In fact, anything above our current warming (or maybe better our pre-industrial temperature) is too much. We have seen this from the recent increase in climate change-related extreme weather events worldwide. With every tenth of a degree of further warming the impacts will be larger and more severely felt. We need everyone and every tool to help us reduce emissions before carbon dioxide becomes a truly wild spirit and puts our environment on a direct path to catastrophe.

BIBLIOGRAPHY

Adam, D., 2010. Climate: The hottest year. Nature, 468(7322), pp.362–364.

Aeschylus, translated by Smyth, H., 1926. Prometheus bound. Cambridge, MA: Harvard University Press.

Agar, J., 2015. 'Future Forecast–Changeable and Probably Getting Worse': The UK Government's Early Response to Anthropogenic Climate Change. Twentieth Century British History, 26(4), pp.602–628.

Agassiz, L., 1840. Études sur les glaciers. Neuchatel: Jent et Gassmann. Available at: https://archive.org/details/etudessurlesgla00agasgoog/page/n9/mode/2up

Alley, R., 2000. The Two-Mile Time Machine: Ice Cores, Abrupt Climate Change, and Our Future. Princeton, NJ: Princeton University Press.

Anderson, T., Hawkins, E., and Jones, P., 2016. CO_2, the greenhouse effect and global warming: from the pioneering work of Arrhenius and Callendar to today's Earth System Models. Endeavour, 40(3), pp.178–187.

Andrew, R., 2020. A comparison of estimates of global carbon dioxide emissions from fossil carbon sources. Earth System Science Data, 12(2), pp.1437–1465.

Arber, A., 1942. Nehemiah Grew (1641–1712) and Marcello Malpighi (1628–1694): An Essay in Comparison. Isis, [online] 34(1), pp.7–16. Available at: https://www.jstor.org/stable/225992.

Archer, D., 2010. The global carbon cycle. Princeton, NJ: Princeton University Press.

Archer, D. and Pierrehumbert, R., 2011. The warming papers. Hoboken: Wiley-Blackwell.

Arora, V., Katavouta, A., Williams, R., Jones, C., Brovkin, V., Friedlingstein, P., Schwinger, J., Bopp, L., Boucher, O., Cadule, P., Chamberlain, M., Christian, J., Delire, C., Fisher, R., Hajima, T., Ilyina, T., Joetzjer, E., Kawamiya, M., Koven, C., Krasting, J., Law, R., Lawrence, D., Lenton, A., Lindsay, K., Pongratz, J., Raddatz, T., Séférian, R., Tachiiri, K., Tjiputra, J., Wiltshire, A., Wu, T., and Ziehn, T., 2020. Carbon–concentration and carbon–climate feedbacks in CMIP6 models and their comparison to CMIP5 models. Biogeosciences, 17(16), pp.4173–4222.

Arrhenius, S., 1896. XXXI. On the influence of carbonic acid in the air upon the temperature of the ground. The London, Edinburgh, and Dublin Philosophical Magazine and Journal of Science, 41(251), pp.237–276.

Arrhenius, S., 1908. Translated by Borns, H., Worlds in the Making: The Evolution of the Universe. New York and London: Harper and Brothers.

Backman, J., Moran, K., McInroy, D., Mayer, L., and the Expedition 302 Scientists, 2006. Expedition 302 summary. Proceedings of the IODP.

Bard, E., 2004. Greenhouse effect and ice ages: historical perspective. Comptes Rendus Geoscience, 336(7–8), pp.603–638.

Bassham, J., Benson, A., Kay, L., Harris, A., Wilson, A., and Calvin, M., 1954. The Path of Carbon in Photosynthesis. XXI. The Cyclic Regeneration of Carbon Dioxide Acceptor. Journal of the American Chemical Society, 76(7), pp.1760–1770.

Berner, R., 1999. A New Look at the Long-term Carbon Cycle. GSA Today, 9(11), pp.1–6.

Berner, R., 2012. Jacques-Joseph Ébelmen, the founder of earth system science. Comptes Rendus Geoscience, 344(11–12), pp.544–548.

Berry, J., Beerling, D., and Franks, P., 2010. Stomata: key players in the earth system, past and present. Current Opinion in Plant Biology, 13(3), pp.232–239.

Betts, R., Jones, C., Liddicoat, S. and Keeling, R., 2022. How the Keeling Curve will need to bend to limit global warming to 1.5 °C. Carbon-Brief. https://www.carbonbrief.org/guest-post-how-the-keeling-curve-will-need-to-bend-to-limit-global-warming-to-1-5c/.

Bolin, B., 2007. A history of the science and politics of climate change. Cambridge: Cambridge University Press.

Bolin, B., Döös, B., Jäger, J., and Warrick, R., 1986. SCOPE 29: The Green-house Effect, Climate Change and Ecosystems. Chichester: J. Wiley and Sons.

Bolin, B. and Erikson, E., 1959. Changes in the carbon dioxide content of the atmosphere and sea due to fossil fuel combustion. The Atmo-sphere and the Sea in Motion—the Rossby Memorial Volume, Journal of Navigation, pp.130–142.

Born, M., 1948. Max Karl Ernst Ludwig Planck, 1858-1947. Obituary Notices of Fellows of the Royal Society, 6(17), pp.161–188.

Bowman, D., Balch, J., Artaxo, P., Bond, W., Carlson, J., Cochrane, M., D'Antonio, C., DeFries, R., Doyle, J., Harrison, S., Johnston, F., Keeley, J., Krawchuk, M., Kull, C., Marston, J., Moritz, M., Prentice, I., Roos, C., Scott, A., Swetnam, T., van der Werf, G., and Pyne, S., 2009. Fire in the Earth System. Science, 324(5926), pp.481–484.

Boyle, R., 1660. New experiments physico-mechanicall, touching the spring of the air, and its effects. Oxford: H. Hall, Printer to the University, for T. Robinson.

Boyle, R., 1661. The Sceptical Chymist or Chymico-Physical Doubts & Paradoxes, Touching the Spagyrist's Principles Commonly call'd Hypo-statical, As they are wont to be Propos'd and Defended by the Generality of Alchymists. Whereunto is præmis'd Part of another Discourse relating to the same Subject. Available at: http://www.gutenberg.org/ebooks/22914

Callendar, G., 1938. The artificial production of carbon dioxide and its influence on temperature. Quarterly Journal of the Royal Meteorological Society, 64(275), pp.223–240.

Calverley, D. and Anderson, K., 2022. Phase out pathways for fossil fuel production within Paris-compliant carbon budgets. [online] Available at: https://www.iisd.org/publications/report/phaseout-pathways-fossil-fuel-production-within-paris-compliant-carbon-budgets.

Carbon Brief. 2015. The most influential climate change papers of all time—Carbon Brief. [online] Available at: https://www.carbonbrief.org/the-most-influential-climate-change-papers-of-all-time.

Charney, J., Arakawa, A., Baker, D., Bolin, B., Dickinson, R., Goody, R., Leith, C., Stommel, H., and Wunsch, C., 1979. Carbon Dioxide and Climate: A Scientific Assessment. Woods Hole, Massachusetts: Climate Research Board, Assembly of Mathematical and Physical Sciences, National Research Council. Available at: https://nap.nationalacademies.org/download/12181.

Choice Reviews Online, 1999. Historical perspectives on climate change. 36(09), pp.36-5140–36-5140.

Ciais, P., Dolman, A., Dargaville, R., Barrie, L., Bombelli, A., Butler, J., Canadell, P., and Moriyama, T., 2010. Geo Carbon Strategy. [online] Rome: Geo Secretariat Geneva. Available at: https://www.globalcarbonproject.org/misc/JournalSummaryGEO.htm.

Committee to Review the Intergovernmental Panel on Climate Change, 2011. Climate change assessments: Review of the processes and

procedures of the IPCC. InterAcademy Council. Available at: https://
www.interacademies.org/sites/default/files/publication/climate_
change_assessments_review_of_the_processes_procedures_of_the_
ipcc.pdf.

Conant, J., 1950. The Overthrow of the Phlogiston Theory: The Chemical
Revolution of 1775–1789. Cambridge, MA: Harvard University Press.

Coordinating Committee on the Ozone Layer, 1986. Related activities to
the work of the coordinating committee on the ozone layer being implemented by UNEP. [online] Nairobi: United Nations Environment Programme. Available at: https://ozone.unep.org/meetings/8th-session-
coordinating-committee-ozone-layer.

Cox, P., Betts, R., Jones, C., Spall, S., and Totterdell, I., 2000. Acceleration
of global warming due to carbon-cycle feedbacks in a coupled climate
model. Nature, 408(6809), pp.184–187.

Crawford, E., 1997. Arrhenius' 1896 Model of the Greenhouse Effect in Context. Ambio, 26(1), pp.6–11. Available at: http://www.jstor.org/stable/
4314543

Crisp, D., Dolman, H., Tanhua, T., McKinley, G., Hauck, J., Eggleston, S.,
and Aich, V., 2021. How Well Do We Understand the Land-Ocean-
Atmosphere Carbon Cycle?

Cruz, R., Harasawa, H., Lal, M., Wu, S., Anokhin, Y., Punsalmaa, B., Honda,
Y., Jafari, M., Li, C., and Huu Ninh, N., 2007: Asia. Climate Change 2007:
Impacts, Adaptation and Vulnerability. Contribution of Working Group
II to the Fourth Assessment Report of the Intergovernmental Panel
on Climate Change, Parry, M., Canziani, O., Palutikof, J., van der Linden, P., and Hanson, C., eds. Cambridge: Cambridge University Press,
pp.469–506. Available at: https://www.ipcc.ch/report/ar4/wg2/.

Cvijanovic, I., Lukovic, J., and Begg, J., 2020. One hundred years of
Milanković cycles. Nature Geoscience, 13(8), pp.524–525.

Dalton, J., 1842. A New System of Chemical Philosophy. 2nd
ed. London: J. Weale. Available at: https://archive.org/details/
newsystemofchemi01dalt/page/n2/mode/2up

Dansgaard, W., 2005. Frozen annals: Greenland ice cap research.
Copenhagen: Niels Bohr Institute.

Desmond, A. and Moore, J., 1991. Darwin. London: Michael Joseph.

Dickson, A., Afghan, J. and Anderson, G., 2003. Reference materials for
oceanic CO_2 analysis: a method for the certification of total alkalinity.
Marine Chemistry, 80(2–3), pp.185–197.

Dietze, E., Theuerkauf, M., Bloom, K., Brauer, A., Dörfler, W., Feeser, I., Feurdean, A., Gedminienė, L., Giesecke, T., Jahns, S., Karpińska-Kołaczek, M., Kołaczek, P., Lamentowicz, M., Latałowa, M., Marcisz, K., Obremska, M., Pędziszewska, A., Poska, A., Rehfeld, K., Stančikaitė, M., Stivrins, N., Święta-Musznicka, J., Szal, M., Vassiljev, J., Veski, S., Wacnik, A., Weisbrodt, D., Wiethold, J., Vannière, B., and Słowiński, M., 2018. Holocene fire activity during low-natural flammability periods reveals scale-dependent cultural human-fire relationships in Europe. Quaternary Science Reviews, 201, pp.44–56.

Dolman, H., 2019. Biogeochemical Cycles and Climate. Oxford: Oxford University Press.

Douglas, J. and Revkin, A., 2001. Bush, in Reversal, Won't Seek Cut In Emissions of Carbon Dioxide. [online] Available at: https://www.nytimes.com/2001/03/14/us/bush-in-reversal-won-t-seek-cut-in-emissions-of-carbon-dioxide.html.

Dufresne, J., 2009. L'effet de serre: sa decouverte, son analyse par la methode des puissances nettes echangées et les effets de ses variations récentes et futures sur le climat terrestre. L'habiliation a diriger des recherches. L'Université Pierre et Marie Curie. Available at: https://www.lmd.jussieu.fr/~jldufres/publi/2009/HDR_JLD.pdf

Earth Negotiations Bulletin, 2009–1. COP 15. Earth Negotiations Bulletin, [online] 12 (459). Available at: https://enb.iisd.org/copenhagen-climate-change-conference-cop15/summary-report.

Earth Negotiations Bulletin, 2009–2. Summary report, 7–19. Earth Negotiations Bulletin, [online] Available at: https://enb.iisd.org/copenhagen-climate-change-conference-cop15/summary-report.

Earth Negotiations Bulletin, 2015. Summary and Analysis. Earth Negotiations Bulletin, [online] 12 (663). Available at: https://enb.iisd.org/download/pdf/enb12663e.pdf.

Earth Negotiations Bulletin, 2019. Earth Negotiations Bulletin, [online] 12 (775). Available at: https://enb.iisd.org/chile-madrid-climate-change-conference-cop25.

Ebelmen, J., 1845. Sur les produits de la décomposition des espèces minérales de la famille des silicates. In Annales des Mines, 7(3), p.66.

Edwards, P., 2010. A vast machine. Cambridge, MA: MIT Press.

Eek, A., 2006. Noordelijke IJszee was ooit warm en zoet. bionieuws, [online] Available at: http://archief.bionieuws.nl/artikel.php?id=2648.

Ekholm, N., 1901. On the Variations of the Climate of the Geological and Historical Past and their Causes. Quarterly Journal of the Royal Meteorological Society, 27(117), pp.1–62.

Emiliani, C., 1955. Pleistocene Temperatures. The Journal of Geology, 63(6), pp.538–578.

Ericson, D., Broecker, W., Kulp, J., and Wollin, G., 1956. Late-Pleistocene Climates and Deep-Sea Sediments. Science, 124(3218), pp.385–389.

Fankhauser, S., Smith, S., Allen, M., Axelsson, K., Hale, T., Hepburn, C., Kendall, J., Khosla, R., Lezaun, J., Mitchell-Larson, E., Obersteiner, M., Rajamani, L., Rickaby, R., Seddon, N., and Wetzer, T., 2021. The meaning of net zero and how to get it right. Nature Climate Change, 12(1), pp.15–21.

Farrington, J., 2000. Achievements in Chemical Oceanography, in 50 years of ocean discovery. Washington, DC: National Academy Press.

Fleming, J., 1998. Historical perspectives on climate change. Oxford: Oxford University Press.

Fonselius, S., Koroleff, F., and Wärme, K., 1956. Carbon Dioxide Variations in the Atmosphere. Tellus, 8(2), pp.176–183.

Foote, E., 1856. Circumstances affecting the heat of the Sun's rays. American Journal of Art and Science, 22, pp.169–194.

Fourier, J., 1824. Translated by Pierrehumbert, R., Mémoire sur les températures du globe terrestre et des espaces planétaires. Oeuvres de Fourier, pp.95–126. Available at: https://geosci.uchicago.edu/%7Ertp1/papers/Fourier1827Trans.pdf

Franz, W., 1997–1. IR-97-034/September, The Development of an International Agenda for Climate Change: Connecting Science to Policy. INTERIM REPORT. Laxenburg, Austria: International Institute for Applied Systems Analysis.

Franz, W., 1997–2. The Development of an International Agenda for Climate Change: Connecting Science to Policy. ENRP discussion paper E97-07.

Friedlingstein, P., O'Sullivan, M., Jones, M., Andrew, R., Hauck, J., Olsen, A., Peters, G., Peters, W., Pongratz, J., Sitch, S., Le Quéré, C., Canadell, J., Ciais, P., Jackson, R., Alin, S., Aragão, L., Arneth, A., Arora, V., Bates, N., Becker, M., Benoit-Cattin, A., Bittig, H., Bopp, L., Bultan, S., Chandra, N., Chevallier, F., Chini, L., Evans, W., Florentie, L., Forster, P., Gasser, T., Gehlen, M., Gilfillan, D., Gkritzalis, T., Gregor, L., Gruber, N., Harris, I., Hartung, K., Haverd, V., Houghton, R., Ilyina, T., Jain, A., Joetzjer, E., Kadono, K., Kato, E., Kitidis, V., Korsbakken, J., Landschützer, P., Lefèvre,

N., Lenton, A., Lienert, S., Liu, Z., Lombardozzi, D., Marland, G., Metzl, N., Munro, D., Nabel, J., Nakaoka, S., Niwa, Y., O'Brien, K., Ono, T., Palmer, P., Pierrot, D., Poulter, B., Resplandy, L., Robertson, E., Rödenbeck, C., Schwinger, J., Séférian, R., Skjelvan, I., Smith, A., Sutton, A., Tanhua, T., Tans, P., Tian, H., Tilbrook, B., van der Werf, G., Vuichard, N., Walker, A., Wanninkhof, R., Watson, A., Willis, D., Wiltshire, A., Yuan, W., Yue, X., and Zaehle, S., 2020. Global Carbon Budget 2020. Earth System Science Data, 12(4), pp.3269–3340.

From, E. and Keeling, C., 1986. Reassessment of late 19th century atmospheric carbon dioxide variations in the air of western Europe and the British Isles based on an unpublished analysis of contemporary air masses by G. S. Callendar. Tellus B, 38B(2), pp.87–105.

Galvez, M. and Gaillardet, J., 2012. Historical constraints on the origins of the carbon cycle concept. Comptes Rendus Geoscience, 344(11–12), pp.549–567.

Gest, H., 1997. A 'misplaced chapter' in the history of photosynthesis research; the second publication (1796) on plant processes by Dr Jan Ingen-Housz, MD, discoverer of photosynthesis. Photosynthesis Research, 53(1), pp.65–72.

Goldschmidt, V., 1934. Drei Vorträge über Geochemie. Geologiska Föreningen i Stockholm Förhandlingar, 56(3), pp.385–427.

Gould, S., 1987. Time's arrow, time's cycle: Myth and Metaphor in the discovery of geological time. Cambridge, MA: Harvard University Press.

Grew, N., 1682. The anatomy of plants with an Idea of a philosophical history of plants, and several other lectures, read before the Royal Society. By Nehemjah Grew ... London: Printed by W. Rawlins, for the author.

Grubb, M., Vrolijk, C., Brack, D., and Forsyth, T., 1999. The Kyoto Protocol, a guide and assessment. Royal Institute of International Affairs, Earthscan publications.

Guerlac, H., 1957. Joseph Black and Fixed Air: a Bicentenary Retrospective, with Some New or Little Known Material. Isis, 48(2), pp.124–151.

Hales, S., 1727. Vegetable staticks, or, An account of some statical experiments on the sap in vegetables. London: W. & J. Innys and T. Woodward. Available at: https://ia802606.us.archive.org/10/items/vegetablestatick00hale/vegetablestatick00hale.pdf

Hays, J., Imbrie, J., and Shackleton, N., 1976. Variations in the Earth's Orbit: Pacemaker of the Ice Ages. Science, 194(4270), pp.1121–1132.

Herschel, W., 1800–1. XIII. Investigation of the powers of the prismatic colours to heat and illuminate objects; with remarks, that prove the different refrangibility of radiant heat. To which is added, an inquiry into the method of viewing the sun advantageously, with telescopes of large apertures and high magnifying powers. Philosophical Transactions of the Royal Society of London, 90, pp.255–283.

Herschel, W., 1800–2. XIV. Experiments on the refrangibility of the invisible rays of the sun. Philosophical Transactions of the Royal Society of London, 90, pp.284–292.

Hershey, D., 1991. Digging Deeper into Helmont's Famous Willow Tree Experiment. The American Biology Teacher, 53(8), pp.458–460.

Heywood, H., 1945. The 1939 Callendar Steam Tables. Nature, 156(3964), pp.462–462.

Hill, R. and Hopkins, F., 1939. Oxygen produced by isolated chloroplasts. Proceedings of the Royal Society of London Series B - Biological Sciences, [online] 127(847), pp.192–210. Available at: https://royalsocietypublishing.org/doi/10.1098/rspb.1939.0017.

Hovi, J., Sprinz, D., and Bang, G., 2010. Why the United States did not become a party to the Kyoto Protocol: German, Norwegian, and US perspectives. European Journal of International Relations, 18(1), pp.129–150.

Hunter, M., 1982. Early problems in professionalizing scientific research: Nehemiah Grew (1641–1712) and the Royal Society, with an unpublished letter to Henry Oldenburg. Notes and Records of the Royal Society of London, 36(2), pp.189–209.

Hutton, J., 1788. X. Theory of the Earth; or an Investigation of the Laws observable in the Composition, Dissolution, and Restoration of Land upon the Globe. Earth and Environmental Science Transactions of The Royal Society of Edinburgh, 1(2), pp.209–304.

Hutton, J., 1795. Theory of the earth: with proofs and illustrations: in four parts. Edinburgh: William Creech.

Imbrie, J. and Palmer-Imbrie, K., 1979. Ice ages: solving the mystery. London: The MacMillan Press.

Ingen-Housz, J., 1779. Experiments upon vegetables. London: Printed for P. Elmsly and H. Payne.

Ingen-Housz, J., 1797. An essay on the food of plants and the renovation of soils. [online] Available at: https://catalog.hathitrust.org/Record/001498167.

International Conference on the Assessment of the Role of Carbon Dioxide and of Other Greenhouse Gases in Climate Variations and Associated Impacts, World Meteorological Organization, United Nations Environment Programme, International Council of Scientific Unions and World Climate Programme, 1986. Report of the International Conference on the Assessment of the Role of Carbon Dioxide and of Other Greenhouse Gases in Climate Variations and Associated Impacts, Villach, Austria, 9–15 October 1985.

IPCC, 1990. First Assessment Report: Climate Change. [online] Geneva: World Meteorological Organization. Available at: https://www.ipcc.ch/reports/?rp=ar1.

IPCC, 1996. Second Assessment Report: Climate Change 1995: The Science of Climate Change. [Houghton, J., Meira Filho, L., Callander, B., Harris, N., Kattenberg, A., and Maskell, K. (eds)]. [online] Cambridge University Press. Available at: https://www.ipcc.ch/report/ar2/wg1/.

IPCC, 2001: Summary for Policy makers. In: Climate Change 2001: The Scientific Basis. Contribution of Working Group I to the Third Assessment Report of the Intergovernmental Panel on Climate Change [Houghton, J.T., Y. Ding, D.J. Griggs, M. Noguer, P.J. van der Linden, X. Dai, K. Maskell, and C.A. Johnson (eds.)]. Cambridge University Press, Cambridge, United Kingdom and New York, NY, USA

IPCC, 2007: Summary for Policymakers. In: Climate Change 2007: The Physical Science Basis. Contribution of Working Group I to the Fourth Assessment Report of the Intergovernmental Panel on Climate Change [Solomon, S., D. Qin, M. Manning, Z. Chen, M. Marquis, K.B. Averyt, M.Tignor and H.L. Miller (eds.)]. Cambridge University Press, Cambridge, United Kingdom and New York, NY, USA.

IPCC, 2013: Summary for Policymakers. In: Climate Change 2013: The Physical Science Basis. Contribution of Working Group I to the Fifth Assessment Report of the Intergovernmental Panel on Climate Change [Stocker, T.F., D. Qin, G.-K. Plattner, M. Tignor, S.K. Allen, J. Boschung, A. Nauels, Y. Xia, V. Bex and P.M. Midgley (eds.)]. Cambridge University Press, Cambridge, United Kingdom and New York, NY, USA.

IPCC, 2018: Global Warming of 1.5 °C. An IPCC Special Report on the impacts of global warming of 1.5 °C above pre-industrial levels and related global greenhouse gas emission pathways, in the context of strengthening the global response to the threat of climate change, sustainable development, and efforts to eradicate poverty

[Masson-Delmotte, V., Zhai, P., Pörtner, H., Roberts, D., Skea, J., Shukla, P., Pirani, A., Moufouma-Okia, W., Péan, C., Pidcock, R., Connors, S., Matthews, J., Chen, Y., Zhou, X., Gomis, M., Lonnoy, E., Maycock, T., Tignor, M., and Waterfield, T. (eds)]

IPCC, 2021–1: Climate Change 2021: The Physical Science Basis. Contribution of Working Group I to the Sixth Assessment Report of the Intergovernmental Panel on Climate Change [Masson-Delmotte, V., Zhai, P., Pirani, A., Connors, S., Péan, C., Berger, S., Caud, N., Chen, Y., Goldfarb, L., Gomis, M., Huang, M., Leitzell, K., Lonnoy, E., Matthews, J., Maycock, T., Waterfield, T., Yelekçi, O., Yu, R., and Zhou, B. (eds)]. Cambridge University Press, Cambridge, United Kingdom and New York, NY, USA

IPCC, 2021–2: Summary for Policymakers. In: Climate Change 2021: The Physical Science Basis. Contribution of Working Group I to the Sixth Assessment Report of the Intergovernmental Panel on Climate Change [Masson-Delmotte, V., Zhai, P., Pirani, A., Connors, S., Péan, C., Berger, S., Caud, N., Chen, Y., Goldfarb, L., Gomis, M., Huang, M., Leitzell, K., Lonnoy, E., Matthews, J., Maycock, T., Waterfield, T., Yelekçi, O., Yu, R., and Zhou, B. (eds)]. Cambridge University Press, Cambridge, United Kingdom and New York, NY, USA, pp.3–32. Available at: https://www.ipcc.ch/report/sixth-assessment-report-working-group-i/

IPCC, 2022–1: Summary for Policymakers [Pörtner, H., Roberts, D., Poloczanska, E., Mintenbeck, K., Tignor, M., Alegría, A., Craig, M., Langsdorf, S., Löschke, S., Möller, V., and Okem, A. (eds)]. In: Climate Change 2022: Impacts, Adaptation, and Vulnerability. Contribution of Working Group II to the Sixth Assessment Report of the Intergovernmental Panel on Climate Change [Pörtner, H., Roberts, D., Tignor, M., Poloczanska, E., Mintenbeck, K., Alegría, A., Craig, M., Langsdorf, S., Löschke, S., Möller, V., Okem, A., and Rama, B. (eds)]. Cambridge University Press. In Press.

IPCC, 2022–2: Summary for Policymakers. In: Climate Change 2022: Mitigation of Climate Change. Contribution of Working Group III to the Sixth Assessment Report of the Intergovernmental Panel on Climate Change [Shukla, P., Skea, J., Slade, R., Al Khourdajie, A., van Diemen, R., McCollum, D., Pathak, M., Some, S., Vyas, P., Fradera, R., Belkacemi, M., Hasija, A., Lisboa, G., Luz, S., and Malley, J. (eds)]. Cambridge and New York: Cambridge University Press.

Isson, T., Planavsky, N., Coogan, L., Stewart, E., Ague, J., Bolton, E., Zhang, S., McKenzie, N., and Kump, L., 2020. Evolution of the Global

Carbon Cycle and Climate Regulation on Earth. Global Biogeochemical Cycles, 34(2). https://doi.org/10.1029/2018GB006061

Jackson, R., 2018. Ascent of John Tyndall. Oxford: Oxford University Press.

Jackson, R., 2019. Eunice Foote, John Tyndall and a question of priority. Notes and Records: the Royal Society Journal of the History of Science, 74(1), pp.105–118.

Jiang, M., Medlyn, B., Drake, J., Duursma, R., Anderson, I., Barton, C., Boer, M., Carrillo, Y., Castañeda-Gómez, L., Collins, L., Crous, K., De Kauwe, M., dos Santos, B., Emmerson, K., Facey, S., Gherlenda, A., Gimeno, T., Hasegawa, S., Johnson, S., Kännaste, A., Macdonald, C., Mahmud, K., Moore, B., Nazaries, L., Neilson, E., Nielsen, U., Niinemets, Ü., Noh, N., Ochoa-Hueso, R., Pathare, V., Pendall, E., Pihlblad, J., Piñeiro, J., Powell, J., Power, S., Reich, P., Renchon, A., Riegler, M., Rinnan, R., Rymer, P., Salomón, R., Singh, B., Smith, B., Tjoelker, M., Walker, J., Wujeska-Klause, A., Yang, J., Zaehle, S., and Ellsworth, D., 2020. The fate of carbon in a mature forest under carbon dioxide enrichment. Nature, 580 (7802), pp.227–231.

Johnson, S., 2009. The invention of air. An experiment, a journey, a new country and the amazing force of scientific discovery. London: Penguin Books Ltd.

Jouzel, J., 2013. A brief history of ice core science over the last 50 yr. Climate of the Past, 9(6), pp.2525–2547.

Kasting, J., 1993. Earth's Early Atmosphere. Science, 259(5097), pp.920–926.

Kaufman, A. and Xiao, S., 2003. High CO2 levels in the Proterozoic atmosphere estimated from analyses of individual microfossils. Nature, 425(6955), pp.279–282.

Keeling, C., 1998. Rewards and Penalties of Monitoring the Earth. Annual Review of Energy and the Environment, 23(1), pp.25–82.

Keeling, R., Piper, S. and Heimann, M., 1996. Global and hemispheric CO_2 sinks deduced from changes in atmospheric O_2 concentration. Nature, 381(6579), pp.218–221.

Keen, S., 2021. The appallingly bad neoclassical economics of climate change. Globalizations, 18(7), pp.1149–1177.

Kennedy, J., 1961. JFK Address at U.N. General Assembly, 25 September 1961. [video] Available at: https://www.jfklibrary.org/asset-viewer/archives/TNC/TNC-20/TNC-20.

Kirchhoff, G., 1860. Ueber das Verhältniss zwischen dem Emissionsvermö-gen und dem Absorptionsvermögen der Körper für Wärme und Licht. Annalen der Physik und Chemie, 185(2), pp.275–301.

Klein, N., 2015. This changes everything: Capitalism vs. The Climate. New York: Simon & Schuster paperbacks.

Köppen, W. and Wegener, A., 1924. Die Klimate der geologischen Vorzeit. Berlin: Gebrüder Borntraeger. Available at: http://www.borntraeger-cramer.de/9783443010881

Kuhn, T., 1970. The structure of scientific revolutions. 2nd ed. Chicago: University of Chicago Press.

Kumar, M., 2008. Quantum, Einstein, Bohr and the great debate about the nature of reality. London: Icon books.

Kuyper, J., Schroeder, H., and Linnér, B., 2018. The Evolution of the UNFCCC. Annual Review of Environment and Resources, 43(1), pp.343–368.

Langway, C., 2008. The history of early polar ice cores. Cold Regions Science and Technology, 52(2), pp.101–117.

Lear, C., Anand, P., Blenkinsop, T., Foster, G., Gagen, M., Hoogakker, B., Larter, R., Lunt, D., McCave, I., McClymont, E., Pancost, R., Rickaby, R., Schultz, D., Summerhayes, C., Williams, C., and Zalasiewicz, J., 2020. Geological Society of London Scientific Statement: what the geological record tells us about our present and future climate. Journal of the Geological Society, 178(1), 239.

Lenton, T. and Watson, A., 2011. Revolutions that Made the Earth. Oxford: Oxford University Press.

Lüthi, D., Le Floch, M., Bereiter, B., Blunier, T., Barnola, J., Siegenthaler, U., Raynaud, D., Jouzel, J., Fischer, H., Kawamura, K., and Stocker, T., 2008. High-resolution carbon dioxide concentration record 650,000–800,000 years before present. Nature, 453(7193), pp.379–382.

Lyell, C., 1830–1833. Principles of geology, being an attempt to explain the former changes of the Earth's surface, by reference to causes now in operation. London: John Murray. 3 Volumes.

Machta, L., 1991. Interview of Lester Machta by Spencer Weart with William Elliott on 1991, April 25. Available at: www.aip.org/history-programs/niels-bohr-library/oral-histories/31417.

Mahon, B., 2003. The Man Who Changed Everything: The life of James Clerk Maxwell. Hoboken, NJ: Wiley.

Mahony, M. and Hulme, M., 2016. Modelling and the Nation: Institutional-ising Climate Prediction in the UK, 1988–92. Minerva, 54(4), pp.445–470.

Malm, A., 2016. Fossil capital: the rise of steam power and the roots of global warming. London: Verso.

Malm, A. and the Zetkin Collective, 2021. White skin, black fuel: On the Danger of Fossil Fascism. London: Verso.

Manabe, S., 1969. Climate and the Ocean Circulation. Monthly Weather Review, 97(11), pp.775–805.

Manabe, S., 1989. Interview of Syukuro Manabe by Spencer Weart on 1989, December 20. Available at: https://www.aip.org/history-programs/niels-bohr-library/oral-histories/5040.

Manabe, S. and Wetherald, R., 1967. Thermal Equilibrium of the Atmosphere with a Given Distribution of Relative Humidity. Journal of the Atmospheric Sciences, 24(3), pp.241–259.

Marcet, A., 1822. XXXIII. Some experiments and researches on the saline contents of sea-water, undertaken with a view to correct and improve its chemical analysis. Philosophical Transactions of the Royal Society of London, 112, pp.448–456.

Marchi, S., Bottke, W., Elkins-Tanton, L., Bierhaus, M., Wuennemann, K., Morbidelli, A., and Kring, D., 2014. Widespread mixing and burial of Earth's Hadean crust by asteroid impacts. Nature, 511(7511), pp.578–582.

Mason, B. and Goldschmidt, V., 1992. Victor Moritz Goldschmidt. San Antonio: Geochemical Society. Available at: https://www.geochemsoc.org/publications/sps/v4vmgoldschmidt

Maxwell, J., 1865. VIII. A dynamical theory of the electromagnetic field. Philosophical Transactions of the Royal Society of London, 155, pp.459–512.

Meinshausen, M., Lewis, J., McGlade, C., Gütschow, J., Nicholls, Z., Burdon, R., Cozzi, L., and Hackmann, B., 2022. Realization of Paris Agreement pledges may limit warming just below 2 °C. Nature, 604(7905), pp.304–309.

Miller, C., 2004. Climate science and the making of a global political order. In: S. Jasanoff, ed., States of Knowledge: The Co-Production of Science and the Social Order. London: Routledge, pp.46–66.

MIT, 1970. Man's impact on the global environment. Assessment and recommendations for action. Report of the Study of Critical Environmental Problems. Cambridge, MA: MIT.

Müller, A., 2014. Viktor Moritz Goldschmidt (1888–1947) and Vladimir Ivanovich Vernadsky (1863–1945): The father and grandfather of geochemistry? Journal of Geochemical Exploration, 147, pp.37–45.

Murphy, N., 2021. Full text of UN Secretary General's Cop26 statement. [online] thenationalnews.com. Available at: https://www.thenationalnews.com/world/cop-26/2021/11/13/full-text-of-un-secretary-generals-cop26-statement/.

Nash, L., 1957. CASE 5. Plants and the Atmosphere. Harvard Case Histories in Experimental Science, Volume II, pp.322–433.

National Research Council, 1979. Carbon Dioxide and Climate: A Scientific Assessment. [online] Available at: https://www.nap.edu/catalog/12181/carbon-dioxide-and-climate-a-scientific-assessment.

United Nations, Secretary General, 2022. Secretary-General Warns of Climate Emergency, Calling Intergovernmental Panel's Report 'a File of Shame', While Saying Leaders 'Are Lying', Fuelling Flames. [online] Available at: https://www.un.org/press/en/2022/sgsm21228.doc.htm.

New York Times, 1988. New York Times, June 24th, 1988 [online] Available at: https://timesmachine.nytimes.com/timesmachine/1988/06/24/issue.html.

Nickelsen, K., 2012. The Path of Carbon in Photosynthesis: How to Discover a Biochemical Pathway. Ambix, 59(3), pp.266–293.

Nickelsen, K., 2015. Explaining photosynthesis: models of biochemical mechanisms 1840–1960. Heidelberg: Springer Verlag.

NobelPrize.org. 2020. Max Planck – Facts. [online] Available at: https://www.nobelprize.org/prizes/physics/1918/planck/facts/

Nordhaus, W., 2013. The climate casino: Risk, uncertainty and economics for a warming world. New Haven, CT: Yale University Press.

O'Brien, C., Huber, M., Thomas, E., Pagani, M., Super, J., Elder, L., and Hull, P., 2020. The enigma of Oligocene climate and global surface temperature evolution. Proceedings of the National Academy of Sciences, 117(41), pp.25302–25309.

Our World in Data. 2022. Our World in Data. [online] Available at: https://ourworldindata.org/.

Pagani, M., Zachos, J., Freeman, K., Tipple, B., and Bohaty, S., 2005. Marked Decline in Atmospheric Carbon Dioxide Concentrations During the Paleogene. Science, 309(5734), pp.600–603.

Pagel, W., 2002. Jean Baptiste Van Helmont. Cambridge: Cambridge University Press.

Pales, J. and Keeling, C., 1965. The concentration of atmospheric carbon dioxide in Hawaii. Journal of Geophysical Research, 70(24), pp.6053–6076.

Palmer, G., 1992. The Earth Summit: What went wrong at Rio? Washington University Law Quarterly, [online] 70 (4). Available at: https://journals.library.wustl.edu/lawreview/article/id/5181/.

Partington, J., 1936. Joan Baptista van Helmont. Annals of Science, 1(4), pp.359–384.

Partington, J., 1960. Joseph Black's 'Lectures on the Elements of Chemistry'. Chymia, 6, pp.27–67.

Peng, W., Iyer, G., Bosetti, V., Chaturvedi, V., Edmonds, J., Fawcett, A., Hallegatte, S., Victor, D., van Vuuren, D., and Weyant, J., 2021. Climate policy models need to get real about people — here's how. Nature, 594(7862), pp.174–176.

Phillips, N., 1956. The general circulation of the atmosphere: A numerical experiment. Quarterly Journal of the Royal Meteorological Society, 82(352), pp.123–164.

Playfair, J., 1802. Illustrations of the Huttonian theory of the earth. Edinburgh: William Creech.

Playfair, J., 1805. Biographical account of the late Dr. James Hutton. Transactions of the Royal Society Edinburgh, 5(3), pp.39–99.

Pongratz, J. and Caldeira, K., 2012. Attribution of atmospheric CO_2 and temperature increases to regions: importance of preindustrial land use change. Environmental Research Letters, 7(3), https://doi.org/10.1088/1748-9326/7/3/034001

Pongratz, J., Reick, C., Raddatz, T. and Claussen, M., 2008. A reconstruction of global agricultural areas and land cover for the last millennium. Global Biogeochemical Cycles, 22(3). https://doi.org/10.1029/2007GB003153

Powell, J., 2015. Four revolutions in the earth sciences. New York: Columbia University Press.

Priestley, J., 1772. XIX. Observations on different kinds of air. Philosophical Transactions of the Royal Society of London, 62, pp.147–264.

Priestley, J., 1775. XXXVIII. An account of further discoveries in air. By the Rev. Joseph Priestley, LL.D. F.R.S. in letter to Sir John Pringle, Bart. P.R.S. and the Rev. Dr. Price, F.R.S. Philosophical Transactions of the Royal Society of London, 65, pp.384–394.

Priestley, J., 1776. Experiments and observations on different kinds of air. London: Printed for J. Johnson. Available at: http://www.gutenberg.org/ebooks/29734

Rae, J., 1895. Life of Adam Smith. New York: Macmillan Publishers.

Rasmussen, D., 2017. The Infidel and the Professor: David Hume, Adam Smith, and the Friendship That Shaped Modern Thought. Princeton, NJ: Princeton University Press.

Rasool, S. and Schneider, S., 1971. Atmospheric Carbon Dioxide and Aerosols: Effects of Large Increases on Global Climate. Science, 173(3992), pp.138–141.

Raupach, M., Marland, G., Ciais, P., Le Quéré, C., Canadell, J., Klepper, G., and Field, C., 2007. Global and regional drivers of accelerating CO_2 emissions. Proceedings of the National Academy of Sciences, 104(24), pp.10288–10293.

Raworth, K., 2017. Doughnut economics. White River Junction: Chelsea Green Publishing.

Reed, E. W., 1992. American women in science before the civil war. [online] Catherinecreed.com. Available at: http://catherinecreed.com/wp-content/uploads/2019/04/women_in_science.pdf

Reick, C., Raddatz, T., Pongratz, J., and Claussen, M., 2010. Contribution of anthropogenic land cover change emissions to pre-industrial atmospheric CO_2. Tellus B: Chemical and Physical Meteorology, 62(5), pp.329–336.

Repcheck, J., 2003. The man who found time: James Hutton and the discovery of the Earth's antiquity. Basic Books: New York.

Revelle, R. and Suess, H., 1957. Carbon Dioxide Exchange Between Atmosphere and Ocean and the Question of an Increase of Atmospheric CO_2 during the Past Decades. Tellus, 9(1), pp.18–27.

Robert, A., 2015. COP21 hailed as shining example of French diplomacy. [online] www.euractiv.com. Available at: https://www.euractiv.com/section/climate-environment/news/cop21-hailed-as-shining-example-of-french-diplomacy/.

Rosen, W., 2010. The most powerful idea in the world. Chicago: University of Chicago Press.

Royer, D., 2014. Atmospheric CO_2 and O_2 During the Phanerozoic: Tools, Patterns, and Impacts. In Treatise on Geochemistry, K. Turekian (ed.), pp.251–267.

Ruddiman, W., 2005. Plows, plagues, and petroleum: how humans tool control of climate. Princeton, NJ: Princeton University Press.

Sagan, C. and Mullen, G., 1972. Earth and Mars: Evolution of Atmospheres and Surface Temperatures. Science, 177(4043), pp.52–56.

Shackleton, N. and Opdyke, N., 1973. Oxygen Isotope and Palaeomagnetic Stratigraphy of Equatorial Pacific Core V28-238: Oxygen Isotope Temperatures and Ice Volumes on a 105 Year and 106 Year Scale. Quaternary Research, 3(1), pp.39–55.

Shapin, S. and Schaffer, S., 2011. Leviathan and the Air-Pump: Hobbes, Boyle and the experimental life. Princeton, NJ: Princeton University Press.

Sharkey, T., 2019. Discovery of the canonical Calvin–Benson cycle. Photosynthesis Research, 140(2), pp.235–252.

Siegenthaler, U., 1984. 19th century measurements of atmospheric CO2 - A comment. Climatic Change, 6(4), pp.409–411.

Sigman, D. and Boyle, E., 2000. Glacial/interglacial variations in atmospheric carbon dioxide. Nature, 407(6806), pp.859–869.

Smil, V., 2017. Energy and civilization. Cambridge, MA: MIT Press.

Speelman, E., van Kempen, M., Barke, J., Brinkhuis, H., Reichart, G., Smolders, A., Roelofs, J., Sangiorgi, F., de Leeuw, J., Lotter, A., and Sinninghe Damsté, J., 2009. The Eocene Arctic Azolla bloom: environmental conditions, productivity and carbon drawdown. Geobiology, 7(2), pp.155–170.

Strathern, P., 2001. Mendeleyev's dream. London: Hamish Hamilton.

Suess, H., 1954. U. S. Geological Survey Radiocarbon Dates I. Science, 120(3117), pp.467–473.

Suess, H., 1955. Radiocarbon Concentration in Modern Wood. Science, 122(3166), pp.415–417.

Supran, G. and Oreskes, N., 2017. Assessing ExxonMobil's climate change communications (1977–2014). Environmental Research Letters, 12(8), p.084019.

Takahashi, T., 1961. Carbon dioxide in the atmosphere and in Atlantic Ocean water. Journal of Geophysical Research, 66(2), pp. 477–494.

Takahashi, T., 1997. Interview of Taro Takahashi by Tanya Levin and Mike Sfraga on 1997, June 27. Available at: https://www.aip.org/history-programs/niels-bohr-library/oral-histories/6995-1

Takahashi, T., Kaiteris, P., and Broecker, W., 1976. A method for shipboard measurement of CO_2 partial pressure in seawater. Earth and Planetary Science Letters, 32(2), pp.451–457.

Tans, P., 2019. Personal communication on 20-9-2019. NOAA.

Thorpe, T., 1906. Joseph Priestley. Nature, 74(1920), pp.378–379.

Tubiana, L., 2021. The French COP 21 Presidency. In Negotiating the Paris Agreement, H. Jepsen, M. Lundgren, K. Monheaim, and H. walker (eds), pp.46–64. Cambridge: Cambridge University Press.

Tyndall, J., 1861. I. The Bakerian Lecture. On the absorption and radiation of heat by gases and vapours, and on the physical connexion of radiation, absorption, and conduction. Philosophical Transactions of the Royal Society of London, 151, pp.1–36.

United Nations, 1992. United Nations Framework Convention on Climate Change FCCC/INFORMAL/84. [online] Available at: https://unfccc.int/resource/docs/convkp/conveng.pdf.

United Nations Development Programme, 2022. Sustainable Development Goals | United Nations Development Programme. [online] UNDP. Available at: https://www.undp.org/sustainable-development-goals.

United Nations Environment Programme, 2020. Emissions Gap Report 2020. [online] UNEP - UN Environment Programme. Available at: https://www.unep.org/emissions-gap-report-2020.

United Nations Environment Programme, 2021. Emissions Gap Report 2021. [online] UNEP - UN Environment Programme. Available at: https://www.unep.org/resources/emissions-gap-report-2021.

United Nations, General Assembly, 1961. General Assembly Resolution 1721 (XVI): International co-operation in the peaceful uses of outer space A/RES/1721. [online] Unoosa.org. Available at: https://www.unoosa.org/oosa/en/ourwork/spacelaw/treaties/resolutions/res_16_1721.html.

United Nations, General Assembly, 1972. Report of the United Nations Conference on the Human Environment, Stockholm, 5–16 June 1972 A/CONF.48/14/Rev.1. [online] United Nations Digital Library System. Available at: https://digitallibrary.un.org/record/523249.

United Nations, General Assembly, 1989. Protection of global climate for present and future generations of mankind. A/RES/43/53. [online] Available at: https://unfccc.int/documents/4588.

United States. President's Science Advisory Committee. Environmental Pollution Panel, 1965. Restoring the Quality of Our Environment: Report. Washington, D.C.: White House.

Urey, H., 1952. On the Early Chemical History of the Earth and the Origin of Life. Proceedings of the National Academy of Sciences, 38(4), pp.351–363.

van Beek, L., Hajer, M., Pelzer, P., van Vuuren, D., and Cassen, C., 2020. Anticipating futures through models: the rise of Integrated Assessment

Modelling in the climate science-policy interface since 1970. Global Environmental Change, 65, p.102191.

van der Gaast, W., 2017. International Climate Negotiation Factors. Cham, Switzerland: Springer.

van Helmont, J-B., 1652. Ortus Medicinæ id est Initia Physicæ Inaudita. Elsevier, Amsterdam, available at https://doi.org/10.1016/C2014-0-02715-3

van Helmont, J-B., 1660. Dageraed, ofte nieuwe opkomst der geneeskonst. Amsterdam: Uitgeverij W.N. Schors. Available as foto reproduction from 1978 https://www.dbnl.org/tekst/helm009dage01_01/colofon.php.

van Niel, C., 1967. The Education of a Microbiologist; Some Reflections. Annual Review of Microbiology, 21(1), pp.1–31.

Vernadsky, V., 1926. The Biosphere. New York: Copernicus.

Vidal, J. and Watts, J., 2009. Copenhagen: The last-ditch drama that saved the deal from collapse. [online] the Guardian. Available at: https://www.theguardian.com/environment/2009/dec/20/copenhagen-climate-global-warming.

Waenke, H. and Arnold, J., 2006. Hans Suess 1909–1993. Biographical Memoirs. National Academy of Sciences. Biographical Memoirs, 87.

Walker, S., Keeling, R., and Piper, S., 2016. Reconstruction of the Mauna Loa Carbon Dioxide Record using High Frequency APC Data from 1958 through 2004. La Jolla, California: Scripps Institution of Oceanography.

Ward, B. and Dubos, R., 1972. Only one Earth: the care and maintenance of a small planet. New York: Norton.

Warde, P. and Marra, A., 2007. Energy consumption in England & Wales, 1560–2000. Consiglio Nazionale delle Ricerche, Istituto di Studi sulle Società del Mediterraneo.

Weart, S., 2003. The discovery of global warming. Cambridge, MA: Harvard University Press.

Wegener, A. and Biram, J., 1966. The origin of continents and oceans. New York: Dover Publications, Inc.

West, J., 2015. Essays on the History of Respiratory Physiology. New York: Springer.

White, J., 2012. Herschel and the Puzzle of Infrared. American Scientist, 100(3), p.218.

Wien, W., 1896. Ueber die Energievertheilung im Emissionsspectrum eines schwarzen Körpers. Annalen der Physik und Chemie, 294(8), pp.662–669.

Williams, M., 2003. Deforesting the Earth. Chicago: University of Chicago Press.

Woodward, J., 2014. The Ice Age. Oxford: Oxford University Press.

Wootton, D., 2015. The invention of science. 1st ed. London: Allen Lane.

World Commission on Environment and Development, 1987. Report of the World Commission on Environment and Development: Our Common Future (a.k.a. the Brundtland Report). [online] United Nations. Available at: https://digitallibrary.un.org/record/139811.

Wright, G., 1898. Agassiz and the Ice Age. The American Naturalist, 32(375), pp.165–171.

Wrigley, E., 2013. Energy and the English Industrial Revolution. Philosophical Transactions of the Royal Society A: Mathematical, Physical and Engineering Sciences, 371(1986), p.20110568.

Zachos, J., Dickens, G., and Zeebe, R., 2008. An early Cenozoic perspective on greenhouse warming and carbon-cycle dynamics. Nature, 451(7176), pp.279–283.

Zahnle, K., Arndt, N., Cockell, C., Halliday, A., Nisbet, E., Selsis, F., and Sleep, N., 2007. Emergence of a Habitable Planet. Space Science Reviews, 129(1–3), pp.35–78.

Zilman, J., 2009. A History of Climate Activities. World Meteorological Organization Bulletin, [online] 58 (3). Available at: https://public.wmo.int/en/bulletin/history-climate-activities.

INDEX